SpringerBriefs in Geography

SpringerBriefs in Geography presents concise summaries of cutting-edge research and practical applications across the fields of physical, environmental and human geography. It publishes compact refereed monographs under the editorial supervision of an international advisory board with the aim to publish 8 to 12 weeks after acceptance. Volumes are compact, 50 to 125 pages, with a clear focus. The series covers a range of content from professional to academic such as: timely reports of state-of-the art analytical techniques, bridges between new research results, snapshots of hot and/or emerging topics, elaborated thesis, literature reviews, and in-depth case studies.

The scope of the series spans the entire field of geography, with a view to significantly advance research. The character of the series is international and multidisciplinary and will include research areas such as: GIS/cartography, remote sensing, geographical education, geospatial analysis, techniques and modeling, landscape/regional and urban planning, economic geography, housing and the built environment, and quantitative geography. Volumes in this series may analyze past, present and/or future trends, as well as their determinants and consequences. Both solicited and unsolicited manuscripts are considered for publication in this series.

SpringerBriefs in Geography will be of interest to a wide range of individuals with interests in physical, environmental and human geography as well as for researchers from allied disciplines.

Fang-Ying Gong

Street Sensing

Urban Thermal Environment Assessment
Using Street View Images

Fang-Ying Gong
School of Public Administration and Policy
Renmin University of China
Beijing, China

ISSN 2211-4165   ISSN 2211-4173 (electronic)
SpringerBriefs in Geography

ISBN 978-3-031-92004-2   ISBN 978-3-031-92005-9 (eBook)
https://doi.org/10.1007/978-3-031-92005-9

© The Editor(s) (if applicable) and The Author(s), under exclusive license to Springer Nature Switzerland AG 2025

This work is subject to copyright. All rights are solely and exclusively licensed by the Publisher, whether the whole or part of the material is concerned, specifically the rights of translation, reprinting, reuse of illustrations, recitation, broadcasting, reproduction on microfilms or in any other physical way, and transmission or information storage and retrieval, electronic adaptation, computer software, or by similar or dissimilar methodology now known or hereafter developed.
The use of general descriptive names, registered names, trademarks, service marks, etc. in this publication does not imply, even in the absence of a specific statement, that such names are exempt from the relevant protective laws and regulations and therefore free for general use.
The publisher, the authors and the editors are safe to assume that the advice and information in this book are believed to be true and accurate at the date of publication. Neither the publisher nor the authors or the editors give a warranty, expressed or implied, with respect to the material contained herein or for any errors or omissions that may have been made. The publisher remains neutral with regard to jurisdictional claims in published maps and institutional affiliations.

This Springer imprint is published by the registered company Springer Nature Switzerland AG
The registered company address is: Gewerbestrasse 11, 6330 Cham, Switzerland

If disposing of this product, please recycle the paper.

Preface

Cities with high-density urban environment, such as Hong Kong, suffer from serious problems related to urban heat island effect, air pollution, and reduced human thermal comfort because city structure drastically alters its micro-climate and thermal environment. To balance the need for providing energy and resources in a high-density environment with the need for reducing urbanization's effects on the environment has been the primary challenges for urban planning. To understand the urban radiation balance and thermal environment, the urban canyon morphology and street-level solar radiation are two crucial factors. Urban canyon morphology, which is defined by different streetscape features (such as sky, tree, and building) and their geometries affect the urban thermal balance. Street-level solar radiation determines the energy input into a street canyon.

However, a large-scale assessment of the impacts of street morphologies and solar irradiation on the street thermal environment is still lacking due to a lack of high-accuracy and high-density measurements. In this book, using publicly available Google Street View (GSV) images and a deep-learning technique, we developed methods for effectively and accurately mapping street sky, tree, and building view factors (VFs) and street-level solar irradiance in a high-density urban environment. Based on the generated maps of street VFs and solar irradiance, we analyzed their spatial and temporal patterns in the high-density urban areas of Hong Kong for the purpose of improving our understandings of the mechanisms of urban energy balance in the high-density urban environment. These understandings will play an important role in providing science-based evidence for urban climatic studies and decision-making in urban planning and design processes.

First, this book presents a methodology for accurately estimating sky view factor (SVF), tree view factor (TVF), and building view factor (BVF) of street canyons in the high-density urban environments using GSV images and a deep-learning algorithm for extraction of street features (sky, trees, and buildings). As a result, SVF, TVF, and BVF maps of street canyons are generated. Verification using reference data of hemispheric photography from field surveys in compact high-rise and low-rise areas shows that the GSV-based VF estimates have a satisfying agreement with the reference data (all with $R^2 > 0.95$), suggesting the effectiveness and high accuracy of the developed method. This is the first reported use of hemispheric photography for direct verification in a GSV-based streetscape study. Furthermore, a comparison between GSV-based and 3D-GIS-based SVFs shows that the two SVF estimates are significantly correlated ($R^2 = 0.40$, $p < 0.01$) and show better agreement in high-density areas. However, the latter overestimates SVF by 0.11 on average, and the differences between them are significantly correlated with street trees ($R^2 = 0.53$): the more street trees, the larger the difference. This suggests that a lack of street trees in a 3D-GIS model of street environments is the dominant factor contributing to the large discrepancies between the two datasets.

Second, this book proposes a method for calculating solar irradiance of street canyons using GSV images and investigates its spatiotemporal patterns in a high-density urban environment. In this method, GSV images provide a unique way to characterize the street morphology from which the diurnal solar path and solar radiation exposure can be estimated in a street canyon. Verifications of our developed method using free-horizon HKO observations and street-level field measurements show that both the calculated clear-sky and all-sky solar irradiance of street canyons well capture the diurnal and seasonal cycles. In the high-density urban areas of Hong Kong, we found that (1) the lowest monthly averaged solar irradiations in winter are 6.6 (December) and 4.6 (February) MJ/m^2/day, and the highest values in summer are 17.3 (July) and 10.8 (June) MJ/m^2/day for clear-sky and all-sky calculations, respectively; (2) The spatial variability of solar irradiation is closely related to sky view factor (SVF). In summer, the irradiation in a low-rise region (SVF ≥ 0.7) on average is about three times that in a high-rise region (SVF ≤ 0.3), and they differ by about five times in winter; (3) Street orientation has a significant impact on the solar radiation received in a high-density street canyon. In general, street canyons with West–East orientation receive higher solar irradiation during summer and lower during winter compared to those with South–North orientation. The generated maps of street-level solar irradiation may help researchers investigate the interactions between solar radiation, human health and urban thermal balance in high-density urban environments.

Finally, this book aims to provide a data-driven and scientific evidence-based strategy for implementations of urban planning and design at street-level in the high-density urban areas of Hong Kong. Easy access to street-level greenery, wide street sky-opening, and adequate exposure to solar radiation have been shown to be important for promoting physical activities and improving public health. With the spatial and temporal mappings of street morphology and street-level solar radiation, street-level hotspots are identified with no street-level greenery, or very small openness, or excessive/insufficient solar exposure. Suggestions on how to improve are also made for informing policymakers what needs to be modified in the practical planning and design stage to mitigate negative effects on the surrounding environments. This research provides a scientific understanding of decision-making during the urban planning and design practices.

Beijing, China   Fang-Ying Gong

Competing Interests The author has no competing interests to declare that are relevant to the content of this manuscript.

Acknowledgements

This book would not have been possible without the guidance, support, and inspiration of many exceptional mentors, colleagues, and loved ones.

First and foremost, I extend my deepest gratitude to my PhD advisor at CUHK, Professor Edward Ng. His mentorship has been instrumental in shaping my academic journey, guiding me toward meaningful and impactful research. His wisdom and encouragement have provided me with a lasting framework for approaching challenges, both in academia and in life. From him, I learned two essential principles: exploration and challenge—that life has no fixed boundaries, only the pursuit of important endeavors, and the continuous journey of self-improvement.

I am also profoundly grateful to my host advisor at MIT, Professor Leslie K. Norford. Our first encounter was during my first year as a PhD student when I served as a Teaching Assistant for his lectures at CUHK. Two years later, I had the privilege of studying and working under his guidance at MIT. His unwavering support and encouragement in both my academic and personal pursuits have left a lasting impact. From him, I learned the importance of responsibility and rigor in academic work—values that I will carry forward in my career.

My sincere appreciation also goes to my advisor at Caltech, Professor Yuk L. Yung, whose insightful suggestions and encouragement helped refine my research. His guidance instilled in me a growth mindset, which will continue to inspire my lifelong learning. The support of my fellow researchers, colleagues, and friends from CUHK, MIT, and Caltech has made this journey even more enriching. Their collaboration and friendship have not only contributed to my research but have also made my academic experience deeply fulfilling.

I am eternally grateful to my beloved parents and husband, whose unwavering love and encouragement have been my greatest source of strength. A special tribute goes to my most beloved grandfather, whose unconditional love and guidance shaped my early years. Though he left my world a decade ago, his influence remains a cherished presence in my heart.

Finally, I dedicate this book to my little daughter and son—a testament to the ever-evolving journey of life. As one chapter concludes, a new path of exploration unfolds. In the words of Winston Churchill: "Now this is not the end. It is not even the beginning of the end. But it is, perhaps, the end of the beginning."

This work was supported by the National Natural Science Foundation of China (Grant No. 42301477) and the Humanities and Social Sciences Youth Foundation, Ministry of Education (Grant No. 22YJCZH034). This research was also supported by the Public Computing Cloud, Renmin University of China (PCC@RUC).

Contents

1	**Issues of Urban Thermal Environment**		1
	1.1	Issues	2
	1.2	Research Background	3
		1.2.1 Effects of Solar Radiation on the Urban Environment	3
		1.2.2 Street View Images in Urban Environmental Research	4
	1.3	Research Questions	6
	1.4	Research Objectives	8
	1.5	Book Structure	10
	References		11
2	**Urban Canyon Morphology and Solar Radiation Dynamics**		15
	2.1	Overview	16
	2.2	Street Canyon Morphology and Quantification	16
		2.2.1 Street View Factors as Morphology Indicators in Urban Climate Study	16
		2.2.2 Methods of Quantifying Street View Factors	18
	2.3	Street Canyon Solar Radiation and Estimations	19
		2.3.1 Physical Basis of Street-Level Solar Irradiance	20
		2.3.2 Methods of Estimating Solar Radiation	23
	2.4	Summary	24
	References		25
3	**Methodological Innovations in Urban Canyon Analysis**		29
	3.1	Overview	30
	3.2	Study Area and Data Collection	30
		3.2.1 Study Area	30
		3.2.2 Data Collection	32
	3.3	GSV-Based Estimation of Street-Level View Factors	35
		3.3.1 Collecting GSV Panorama Images	35

		3.3.2	Extractions of Street Features Using Deep-Learning Techniques	37
		3.3.3	Projection Into Fisheye Images and Calculations of View Factors	38
	3.4	GSV-Based Estimation of Street-Level Solar Radiation		39
		3.4.1	Attributes Collection and Features Extractions from GSV Images	40
		3.4.2	Urban Canyon Geometry Calculation Using GSV Images and Solar Path	42
		3.4.3	Calculation of Street-Level Solar Radiation	43
	3.5	Summary		48
	References			49
4	**Spatial Patterns of Street Canyon View Factors**			51
	4.1	Overview		52
	4.2	Methods Verification		52
		4.2.1	Accuracy Assessment of Features Classification by Deep Learning	52
		4.2.2	Verification of GSV-Based View Factor Estimates	54
		4.2.3	3D-GIS-Based SVF Estimates	55
	4.3	Results		57
		4.3.1	Mapping SVF, TVF, and BVF of Street Canyons Using GSV Images	57
		4.3.2	Comparison between GSV-Based and 3D-GIS-Based SVF Estimates	59
		4.3.3	Impacts of Street Tree Canopy and Building Density on SVF Estimates	59
	4.4	Discussion		62
		4.4.1	Large Uncertainty in Model-Based SVF Estimates from Street Trees	62
		4.4.2	Temporal Variation of Street-Level View Factors	63
		4.4.3	Spatial Variation of Street-Level View Factors	65
	4.5	Summary		65
	References			66
5	**Spatiotemporal Patterns of Street Canyon Solar Radiation**			69
	5.1	Overview		70
	5.2	Methods Verification		71
		5.2.1	Accuracy Assessment of GSV-Based Solar Radiation Method	71
		5.2.2	Verification of GSV-Based Street-Level Solar Radiation Estimates	73
		5.2.3	Verification of GSV-Based Free-Horizon Solar Radiation Estimates	74
	5.3	Results		77

		5.3.1	Spatiotemporal Pattern of Clear-Sky Street-Level Solar Irradiation	77
		5.3.2	Spatiotemporal Pattern of all-Sky Street-Level Solar Irradiation	78
		5.3.3	Contributions from Direct and Diffuse Components	81
		5.3.4	Effect of Street Canyon Geometry on Solar Irradiation	82
	5.4	Discussion		85
		5.4.1	Spatial Inhomogeneity of Solar Radiation	85
		5.4.2	Reflected Radiation in a Street Canyon and its Impact	86
		5.4.3	Transmissivity of Solar Radiation Through Tree Crowns	89
	5.5	Summary		89
	References			90
6	**Implementation of Urban Environmental Planning and Governance at Street Level**			93
	6.1	Overview		94
	6.2	Implications on Street-Level Greenery		95
		6.2.1	Significance of TVF Map for Urban Planning Practices	95
		6.2.2	Hotspots Without Street Greenery in Hong Kong	96
		6.2.3	Improving Street Greenery in Hong Kong	98
	6.3	Implications on Street-Level Sky Openness		99
		6.3.1	Significances of SVF Map for Urban Planning Practices	99
		6.3.2	Hotspots with Very Low Street Openness in Hong Kong	100
		6.3.3	Improving Street Openness in Hong Kong	102
	6.4	Implications on Street-Level Solar Exposure		103
		6.4.1	Significances of Street Solar Radiation Estimates for Urban Planning	103
		6.4.2	Solar Under-Exposure in Winter and Over-Exposure in Summer	104
		6.4.3	Recommendations for Urban Street Planning and Design	106
	6.5	Potential Applications on Urban Climatic Study and Urban Planning Practices		106
		6.5.1	Verification of Urban Morphology and Microclimate Models	106
		6.5.2	Assessment of the Feasibility for Installing Solar Panels	107
		6.5.3	Estimation of Effective Albedo Based on View Factors	108
		6.5.4	Impacts of Street Geometries on Urban Microclimate	108
		6.5.5	Urban Planning Practices for Old and New Town Developments	109

	6.5.6	Comparison Analysis of Global High-Density Cities	112
6.6	Summary		114
References			114

7 Advancements and Future Directions in Urban Street Sensing Methodologies ... 117

7.1	Summary of Contributions		118
7.2	Strengths of Street Sensing Method		120
7.3	Assumptions of Street Sensing Method		120
	7.3.1	Spatial and Temporal Variations of Street View Factors	120
	7.3.2	Spatial Inhomogeneity of Solar Radiation	121
	7.3.3	Transmissivity of Solar Radiation Through Tree Crowns	121
	7.3.4	Reflected Radiation and Its Impacts	122
	7.3.5	Corrections for Global Measurements Under Cloudy Skies	122
	7.3.6	Impact of Sky Luminance Distribution on Diffuse Radiation Estimation	123
7.4	Limitations and Future Works		124
References			125

Index ... 127

List of Figures

Fig. 2.1 The physical streetscape using view factors, including the main street components, sky, trees, buildings and ground in a high-density environment. 20

Fig. 2.2 Physical basis of incident solar radiation at street level in a high-density environment. As indicated, the street-level solar radiation incident on a surface includes direct and diffuse components from atmospheric molecules (Rayleigh scattering) and aerosols (Mie scattering), and clouds (Liou, 2002; Ross, 1981), and the reflected radiation from buildings and ground. The radiation can be blocked by obstructions, such as buildings and trees. The transmission of sunlight is subject to atmospheric absorptions which can be estimated by surface atmospheric pressure and Linke turbidity factors. 21

Fig. 3.1 (**a**) Location of Hong Kong (yellow circle) in southeastern China; (**b**) High-density urban areas in Hong Kong, as outlined in yellow, including Kowloon and northern Hong Kong Island; the yellow points are the spatial destruction of building footprints; the red points are the locations of King's Park and Kau Sai Chow sites from Hong Kong Observatory; the black point is the location of field measurement of in our study (one street canyon of the campus at the urban area of New Territories); (**c**) building density map, including distribution and height, overlaid with streets in gray. The building and street data are extracted from the B5000 maps series by the Hong Kong Lands Department. The dotted red and blue rectangles outline the field survey regions for high-rise and low-rise regions, respectively, as described in Sect. 4.2.2. The black star is the location of the street canyon example in Mong Kok shown in Fig. 3.2 31

Fig. 3.2	An example of the deep street canyon in the Mong Kok area (the black star shown in Fig. 3.1), one of the typical high-density high-rise urban areas of Hong Kong (Google Street View, 2017)......	32
Fig. 3.3	Google Street View coverage 3-D map of Kowloon Area (Google Street View, 2017)	33
Fig. 3.4	Monthly averaged Linke turbidity factors (in black) in Hong Kong. Based on Table 2 in Li and Lam (2002), the Linke turbidity factor used in this study is the average value of three different estimates: T_{Lin} (in red), T_{Lou} (in green), and T_{Pin} (in blue) ...	35
Fig. 3.5	Workflow procedure for VF calculations using GSV images, illustrated by taking two examples from high-rise and low-rise areas. (a) Panorama images downloaded from Google servers using coordinates of sampling street points as inputs. (b) Extraction of sky (in blue), trees (in green), and buildings (in gray) using the scene parsing deep-learning technique (Zhao et al., 2016). (c) Fisheye images obtained by projecting the panorama images from cylindrical projection to azimuthal projection. Based on the fisheye image of extracted features, SVF, TVF, and BVF are calculated using the classical photographic method developed by Johnson and Watson (1984). The resulted VF estimates are also indicated	36
Fig. 3.6	Workflow of semantic scene parsing using PSPNet. For a given input street view image in (a), the network extracts the feature map in (b), and then the pyramid parsing module is applied to form the final feature representation of the streetscape in (c). Finally, a pixel-wise classified output street view image with semantic categories in (d) produced by feeding the feature representation into a convolution layer...........................	38
Fig. 3.7	Schematic framework for this study, in which black rectangles represent the collected datasets, green rectangles represent the calculations of solar path and view factors in street canyons using GSV images, the blue rectangle represents the calculation of clear-sky street-level solar irradiance, and the red rectangle represents the calculation of all-sky street-level solar irradiance in the high-density urban areas of Hong Kong	40
Fig. 3.8	Workflow procedure for solar radiation calculation using GSV image shown using an example of a street canyon in high-density urban areas of Hong Kong. (a) Panorama image collected using the GSV API, and extractions of sky (in blue), trees (in green), and buildings (in deep gray) using a deep-learning technique. (b) Street geometries, including zenith angle indicated by concentric circles and azimuth angle indicated by radius lines, in a fisheye and the overlaid sun paths of summer (August 1st) and winter (January 1st); (c) Same as (b) but with classified fisheye images. In addition, solar hour and clear-sky solar radiation are also indicated............................	41

List of Figures   xvii

Fig. 3.9   (a) The direct radiation, and isotropic and anisotropic components of the diffuse radiation incident on the street surface. (b) The Rayleigh scattering from atmospheric molecules and Mie scattering from large atmospheric particles 47

Fig. 3.10   The change of isotropic and anisotropic components of diffuse radiation with solar zenith angle from 0 to 90 under free-horizon condition 48

Fig. 4.1   Spatial distribution of the 100 randomly selected samples in the study area for verification. 53

Fig. 4.2   Accuracy assessment of feature extraction using the PSPNet in a deep-learning framework to calculate SVF, TVF, and BVF from GSV images in high-density urban areas of Hong Kong. The R^2 and RMSE between the two datasets of VFs are also indicated 54

Fig. 4.3   Examples of fisheye images from two high-rise and two low-rise street sample points from a field survey in (a), and GSV-based method in (b). Image features are classified into the sky (in blue), tree (in green), and building (in gray) using the scene parsing deep-learning technique, as shown in (c). SVF, TVF, and BVF values from field surveys and GSV are shown as indicated 55

Fig. 4.4   (a) Scatter plot of SVF reference data from field survey and the corresponding GSV-based (in blue) and 3D-GIS-based (in red) SVF estimates. Sampling SVF data include 20 samples in Mong Kok within high-rise building area (in triangles), and 20 samples in Kowloon Tong within the low-rise area (in circles); (b) the same as (a) but for TVF; (c) the same as (a) but for BVF 56

Fig. 4.5   Maps of GSV-based SVF in (a), TVF in (b), and BVF in (c) of street canyons in high-density urban areas of Hong Kong derived from 29,264 GSV images along streets at 30-m intervals; (d) Frequency density of SVF (blue bar), TVF (green bar), and BVF (gray bar) 58

Fig. 4.6   (a) Map of 3D-GIS-based SVF estimate with the same street sampling points as shown in Fig. 4.5a; (b) Map of the difference between GSV-based and 3D-GIS-based SVF estimates; (c) Bivariate histogram of GSV-based and 3D-GIS-based SVF estimates of street canyon in high-density urban areas of Hong Kong as shown in Figs. 4.5a and 4.6a, respectively. To make the histogram, the SVF data from both datasets are grouped into 0.01×0.01 grids and the value of a grid is the total number of SVF samples that fall in this grid; (d) Comparison of frequency density histogram from GSV-based and 3D-GIS-based SVF estimates 60

Fig. 4.7 (a) Bivariate histogram of GSV-based TVF and the error of SVF calculations from 3D-GIS, quantified using the difference between 3D-GIS-based and GSV-based SVFs. As indicated, when TVF > 0.1, the R^2 is 0.53 and the best fit linear slope is 1.17 (in dotted black line); (b) Bivariate histogram of GSV-based BVF and the error of SVF calculations from 3D-GIS. To derive the histogram, the data from both datasets are grouped into 0.01× 0.01 grids and the value for each grid is the total number of SVF samples that fall in the grid. The color shading indicates the number density and redder color indicates more data points at the grid. These two figures are used to quantify the impacts of TVF and BVF on the errors of SVF calculations from 3D-GIS 61

Fig. 4.8 Spatial distribution of the acquisition time, including year in (a) and season in (b), of the GSV images collected in the high-density urban area of Hong Kong in this study 64

Fig. 5.1 The free-horizon solar irradiance measurements by HKO for 22 May 2018 in (a) and 23 May 2018 in (b) 72

Fig. 5.2 King's Park and Kau Sai Chau meteorological stations. (a) Solar radiation measuring equipment at the King's Park (KP) HKO Station (Hong Kong Observatory, 2003c). (b) Direct and diffuse solar radiation instruments (left) mounted on a sun tracker and the global solar radiation sensor (right) in Kau Sai Chau solar station. 72

Fig. 5.3 Comparison of measured and GSV-based estimated global irradiance in the street canyon measurement site located in the campus of The Chinese University of Hong Kong. (a) Fisheye image of the street canyon; (b) Sky, buildings, and trees features of the street canyon; Comparison of measured and GSV-based estimated global irradiance between 05:00 h and 20:00 h for 22 May 2018 in (c) and 23 May 2018 in (d); Scatter plots between GSV-based estimated and field measured global irradiance from 5:00 h to 20:00 h for 22 May 2018 in (e) and 23 May 2018 in (f) 74

Fig. 5.4 The histograms of hourly averaged sea-level pressure in (a) and cloudiness in (b) between 8:00 h and 18:00 h from HKO measurements from 2009 to 2014 75

Fig. 5.5 Scatter plots between the calculated free-horizon hourly solar irradiance and HKO measurements at the site of King's Park. The plots are grouped by hourly solar irradiation, including direct and diffuse components, under different cloud coverages in terms of octas as measured for 6 years by HKO from 2019 to 2014 75

List of Figures

Fig. 5.6 Comparison of estimated all-sky irradiance and HKO measurement of hourly global, direct, and diffuse solar irradiance for three different cloudiness example days: (**a**) A clear day (octas = 0–1); (**b**) A semi-cloudy day (octas = 2–6); (**c**) A cloudy day (octas = 7–8) 76

Fig. 5.7 Monthly mean of daily clear-sky solar irradiation (MJ/m^2/day) in street canyons averaged over 6 years from 2009 to 2014 in the high-density urban areas of Hong Kong. Calculation of the clear-sky solar irradiation is introduced in Sect. 3.4.3............. 78

Fig. 5.8 Same as Fig. 5.7 but for all-sky street-level solar irradiation. These monthly means of daily all-sky solar irradiation (MJ/m^2/day) in street canyons are averaged over 6 years from 2009 to 2014 in the high-density urban areas of Hong Kong. The detail of the calculation is introduced in Sect. 3.4.3 79

Fig. 5.9 Comparative analysis of monthly mean of the calculated daily clear-sky and all-sky solar irradiation (MJ/m^2/day) based on 25,654 street canyon samples using Google Street View images in the high-density urban area of Hong Kong 80

Fig. 5.10 (**a**) Daily direct irradiation (MJ/m^2/day) of street canyons in high-density urban areas of Hong Kong in January averaged for 6 years (2009–2014); (**b**) Daily diffuse irradiation (MJ/m^2/day) of street canyons in January averaged for 6 years (2009–2014); (**c**) the same as (**a**) but for July, an example of summer; (**d**) The same as (**b**) but for July. These are all-sky street-level solar irradiation estimates based on GSV images and HKO measurements as described in Sect. 3.4.3 82

Fig. 5.11 Coefficient of variation, defined as the ratio of standard deviation to mean, of daily solar irradiation calculated using 2190 days of all-sky solar irradiation data over the 6 years from 2009 to 2014 83

Fig. 5.12 Six different types of street canyons, including three different types of street geometries with H/W ratios of 1/2, 1, and 2, respectively, and two different types of street orientations of North–South and West–East. The daily all-sky solar irradiation (MJ/m^2/day) for summer and winter are also indicated, respectively, as well as sky view factors (SVF), tree view factors (TVF), and building view factor (BVF)........................ 84

Fig. 5.13 Monthly mean of daily all-sky solar irradiation (MJ/m^2): (**a**) global irradiation, (**b**) diffuse irradiation and (**c**) direct irradiation for the six types of street canyons in high-density urban areas of Hong Kong averaged for 6 years from 2009 to 2014....... 85

Fig. 5.14	(a) Comparison of monthly mean of daily solar irradiation measured at King's Park (KP) site and Kau Sai Chau (KSC) site of Hong Kong Observatory from 2009 to 2013; (b) The histogram of the difference between daily solar irradiation at KP and KSC.	87
Fig. 5.15	Monthly mean of daily clear-sky solar hours in street canyons, averaged over 6 years from 2009 to 2014 in the high-density urban areas of Hong Kong, in the winter in (a) and the summer in (b). Comparison analysis of the frequency of solar hours in summer (July) and in winter (January) in (c).	88
Fig. 6.1	The population density of high-density urban areas in Hong Kong based on Street Block level, including eight District Councils with five in Kowloon: Kowloon City (KC), Yau Tsim Mong (YTM), Sham Shui Po (SSP), Kwun Tong (KT), and Wong Tai Sin (WTS), and three in Hong Kong Island: Central & Western (CW), Wan Chai (WC), and Eastern (ET). The statistics of demographic are publicly available from the 2011 census in Hong Kong (Census and Statistics Department, 2016)	95
Fig. 6.2	Street hotspots without greenery (TVF = 0) in high-density urban areas of Hong Kong. The orange dots are regions with TVF equals zero, indicating no street greenery in this region. Eight examples with their fisheye images from Google Street View (with their IDs) are also shown.	97
Fig. 6.3	Identifying the hotspots of very low sky openness (SVF ≤ 0.2). In particular, the green dots are regions where high tree density (TVF ≥ 0.3) block the sky openness; and the red dots are regions where high building density block the sky openness	101
Fig. 6.4	Projecting obstructions: (a) Vertical overhanging signboards is benefited for increasing more openness (increasing SVF) compared with horizontal type; (b) Horizontal overhanging signboards should be avoided. Two realistic fisheye images and its SVF values are described.	102
Fig. 6.5	The distribution of the urban green parks of different scales in Hong Kong. The distribution of existing urban open space, 447 patches, in the study area; The three areas marked by black frames at Mong Kok (MK), Central and Western (CW), and Wan Chai (WC) in high-density areas of Hong Kong have very low street sky openness and lack neighborhood open spaces	103

List of Figures

Fig. 6.6 (**a**) Identifying the street points of potential insufficient solar exposure in Winter (Jan.). The blue and green dots are regions at street level with daily solar radiation less than 2.0 MJ/m^2. This threshold is the lowest quarter of the maximum solar radiation which is about 8.0 MJ/m^2. In particular, the blue dots are regions with high building density while the green dots are regions with high tree density. (**b**) Identifying the points of potential excessive solar exposure in Summer (Jul.). The orange dots are regions at street level with daily solar radiation more than 15.0 MJ/m^2. This threshold is the highest quarter of the maximum solar radiation which is about 20.0 MJ/m^2 105

Fig. 6.7 The spatial distribution of streets, buildings, and normalized difference vegetation index (NDVI) in Mong Kok area. The (NDVI) at 1.2 m resolution is calculated using Worldview-3 data. Different vegetation type with different NDVI is also shown from grassland (light green) to trees (deep green). The five themed streets: Flower Market Road, Tung Choi Street, Sai Yee Street, Fa Yuen Street and Nelson Street from the Mong Kok Revitalisation Project are also indicated 110

Fig. 6.8 (**a**) Map of SVF in Mong Kok and (**b**) map of TVF in Mong Kok area. The dotted rectangle highlights the area without street greening and these streets have a strong need for greening improvement. The street names are also indicated 111

Fig. 6.9 The spatial distribution of street view factors estimates derived from 63, 488 Google Street View panoramas every 30-m interval in Singapore, including SVF in (a_1), TVF in (b_1), and BVF in (c_1); Selected areas in central Singapore [black frame in figure (**a**), (**b**), and (**c**)] in (a_2), (b_2), and (c_2) 113

List of Tables

Table 3.1	Summary of input parameters, data source, the temporal resolution used in this study, and corresponding dataset descriptions	34
Table 5.1	Summary of time information and site information of street canyon field measurement, and associated atmospheric conditions used in the model	71
Table 6.1	Greenery improvement strategies for the three main types of street canyons in high-density urban areas in Hong Kong. Both types do not have street greenery and roadside trees are not encouraged because they may block the sightline of pedestrians and drivers on streets	99

About the Author

Fang-Ying Gong is currently an Assistant Professor at the School of Public Administration and Policy, Renmin University of China, and a Distinguished Young Scholar of the RUC. She completed her Ph.D. in Architecture at the Chinese University of Hong Kong from 2014 to 2019, with joint training in building technology at the School of Architecture and Planning, Massachusetts Institute of Technology (MIT) from 2017 to 2018. She served as a Research Associate at the Department of Earth and Planetary Sciences, California Institute of Technology (Caltech), 2020–2021. Her research focuses on smart cities and spatial governance, resilient cities and environmental governance, as well as digital government and data governance. Her work has been awarded the Postgraduate Research Output Award from The Chinese University of Hong Kong (2018), the Global Excellence Research Award (2017), and the First Prize in Humanities and Social Sciences Research in Macau (2012). As a Principal Investigator, Dr. Gong has led projects funded by the National Natural Science Foundation of China, the Ministry of Education's Social Science Fund, and the Fundamental Research Funds for Central Universities. She has also been a key researcher in international projects with NASA-JPL, the Macao SAR Government, and the Hong Kong Planning Department. Her work has contributed significantly to climate-adaptive urban planning and resilient urban governance.

Nomenclature

Symbols

x_p, y_p	Coordinates of the cylindrical panorama
x_f, y_f	Coordinates of the fisheye image
C_x, C_y	Coordinates of the center pixel on the fisheye image
H_p	Height of the panorama image (mm)
W_p	Width of the panorama image (mm)
R^2	Coefficient of determination
R	Coefficient of correlation
p-value	Significant level
$\alpha_{i,\,x}$	Angular width of pixels of feature x (x can be sky, tree, or building) in the ith ring
r_0	Radius of the fisheye image
T_{mrt}	Mean radiant temperature (°C)
ΔT_{u_r}	Urban heat island intensity (°C)
G	Global solar radiation flux on horizontal surface (W/m²)
G_0	Global solar radiation for undisturbed conditions (free horizon and no clouds) (W/m²)
I	Direct solar radiation flux on horizontal surface (W/m²)
I_0	Extraterrestrial solar radiation (W/m²)
D	Diffuse solar radiation flux on horizontal surface (W/m²)
D_{iso}	Isotropic diffuse radiation (no clouds) (W/m²)
D_{aniso}	Anisotropic diffuse radiation (no clouds) (W/m²)
Q	Heat energy (W)
Q^*	Net all-wave radiation flux density (W/m²)
p	Local atmospheric pressure (hPa)
p_0	Normal pressure at sea level (hPa)
m_{r_0}	Optical air mass
T_L	Linke turbidity factor

Greek Symbols

φ Solar zenith angle (degree)
ψ Solar azimuth angle (degree)
α Surface albedo
λ_f Frontal area index
Ψ_x View factor for sky, tree, and building when x is specified
ε Surface emissivity
τ Transmittance of the direct solar radiation
δ Solar declination angle
δ_{r_0} Vertical optical thickness of the standard atmosphere
γ Solar altitude angle (degree)
θ Polar angle (degree)

Abbreviations

3D-GIS	Three-dimensional geographic information system
API	Application programming interface
AWSs	Automatic weather stations
BVF	Building view factor
C&SD	Hong Kong Census and Statistics Department
CEDD	Civil Engineering and Development Department
CNN	Convolutional neural network
CW	Central and Western
DCs	District councils
DEM	Digital elevation model
DHI	Diffuse horizontal irradiation
DNI	Direct normal irradiation
DEM	Digital elevation model
DSM	Digital surface model
FCN	Fully convolutional network
FVC	Fractional vegetation cover
GHI	Global horizontal irradiation
GIS	Geographic information system
GMP	Greening master plan
GSV	Google Street View
GTI	Global irradiation at optimum tilt
H/W	Building-height-to-street-width ratio
HKEPD	Environmental Protection Department of Hong Kong
HKO	Hong Kong Observatory
HKPSG	Hong Kong Planning Standards and Guidelines
IPCC	Intergovernmental Panel on Climate Change

KC	Kowloon City
KP	King's Park
KSC	Kau Sai Chow
KT	Kwun Tong
LandsD	Hong Kong Lands Department
LST	Land surface temperature
LU/LC	Land use and land cover
MK	Mong Kok
MRT	Mean radiant temperature
NDVI	Normalized difference vegetation index
Plan D	Hong Kong Planning Department
PSPNet	Pyramid scene parsing network
RMSE	Root-mean-square error
SA	Surface albedo
SBs	Street blocks
SSP	Sham Shui Po
SVF	Sky view factor
SZA	Solar zenith angle
TCR	Tree cover ratio
TPUs	Tertiary planning units
TVF	Tree view factor
UBL	Urban boundary layer
UCL	Urban canopy layer
UHI	Urban heat island
UHII	Urban heat island intensity
UNFPA	United Nations Population Fund
URL	Uniform resource locator
VF	View factor
WC	Wan Chai
WTS	Wong Tai Sin
YTM	Yau Tsim Mong

Chapter 1
Issues of Urban Thermal Environment

Contents

1.1	Issues	2
1.2	Research Background	3
	1.2.1 Effects of Solar Radiation on the Urban Environment	3
	1.2.2 Street View Images in Urban Environmental Research	4
1.3	Research Questions	6
1.4	Research Objectives	8
1.5	Book Structure	10
References		11

Abstract Urbanization is rapidly reshaping global cities, influencing local climates, and altering urban thermal environments. This chapter explores the role of urban morphology and street-level solar radiation in regulating urban climate, with a particular focus on high-density environments. Compact urban structures impact energy consumption, human comfort, and ecological balance, making precise assessment crucial for sustainable urban planning. Using Google Street View (GSV) imagery, this study develops a cost-effective and scalable method to quantify urban street canyon morphology and solar radiation. It integrates machine learning and spatial analysis to extract sky openness, tree canopy, and building density, offering a refined approach to urban thermal assessment. The findings enhance our understanding of street-level solar exposure and its implications for thermal comfort, health risks, and energy efficiency. The study provides valuable insights for urban policymakers, aiding the development of climate-responsive planning strategies to mitigate urban heat stress while ensuring adequate sunlight access in high-density cities.

Keywords Urban thermal environment · Street view imagery · Solar radiation · Urban morphology · High-density cities

1.1 Issues

Our earth has been becoming an urban planet. More than half of the world's people now live in cities, and the proportion is growing rapidly. By 2030, there will be over 170 cities in the world with a population over a million, according to projections by the United Nations Population Division (UNFPA, 2017). Particularly, nearly all future population growth will be in the developing countries, especially in Africa and Asia. As the urban areas are sprawling fast, it is becoming more challenging to balance the need for providing energy and resources for growing populations with the need for reducing urbanization's effects on the environment. Sprawling urban areas create environmental problems, including air pollution, urban heat island effect, traffic noise and artificial light, and change the ecosystem in urban due to the spread of impervious surfaces such as paved roads and roofs. The process of urbanization brings a huge impact on the local climate by altering the urban thermal environment and modifying the energy balance in the urban system.

Compact living in cities renders the efficiency of cities in the use of natural resources by decreasing the land use and energy consumption per person. However, the serious environmental problems caused by compact building blocks in a high-density urban environment must be addressed by planning and design in a scientific way. Better environmentally sensitive urban planning and design strategies are evidently necessary to make compact living more valid and practical. The urban thermal environment has practical implications for energy consumption, human comfort and productivity, air pollution at the street level, and urban ecology (Brager & de Dear, 1998; Saito et al., 1990). Among all the factors contributing to the changing urban thermal environment, urban morphologies and solar irradiation of the street canyons are two of the most important ones.

Urban morphology characterizes the structure and geometry of a street canyon, including building blocks, sky opening, street trees, and impervious ground covers (Johansson, 2006). It has been shown that in numerical simulation, if the urban morphological information is more accurate, simulations for air pollution and climate change scenarios can achieve a better result. Therefore, to improve our understandings of the impact of urban morphological characteristics on urban climate, it is essential to obtain high-quality urban morphological information.

Solar radiation is the main driver in regulating urban climate and the street-level thermal energy balance (Oke, 1988). It has been extensively investigated in different fields including urban meteorology (Oleson, 2011; Sanusi et al., 2016), photovoltaic generation (Jakubiec & Reinhart, 2013), urban heat island effect (Oke, 1982), and such related issues as thermal comfort and human health due to UV exposure (Farrar et al., 2013; Rhodes et al., 2010; Webb, 2006). Due to the increasing trend in urbanization and an expected warmer climate in the near future, more and more residents are prone to heat stress in cities (Akbari et al., 2001; Thorsson et al., 2010). Therefore, better quantification of solar irradiance at the urban street level will greatly improve our understanding of the interactions between radiation, human health and the urban thermal environment.

Therefore, an effective and high-accurate method for quantifying the urban morphologies and solar radiation of the street canyons is therefore crucial for studying its urban climate and assessing the relevant outdoor thermal comfort. However, a costly-effective and accurate method for mapping the street canyon morphologies and solar radiation at large scale is still not available.

1.2 Research Background

1.2.1 Effects of Solar Radiation on the Urban Environment

Urban street solar radiation is not only a crucial driving factor for the urban thermal environment, but the level of sunlight exposure also directly impacts the residential quality and physical and mental health of urban residents. With the rapid development of China's urbanization process, the urbanization rate is projected to increase from 63.89% in 2020 to 75.27% in 2035 (Institute of Population and Labor Economics, Chinese Academy of Social Sciences, 2021). Particularly, the rapid expansion of high population density and high building density cities in recent years has exacerbated issues such as urban heat island effects and a decline in residential quality and residents' physical and mental health due to a lack of sunlight exposure (Zhao et al., 2021; Zielinska-Dabkowska & Xavia, 2019). Therefore, precise quantification and assessment of urban street canyon solar radiation can provide important scientific evidence for improving the urban heat environment and reducing health risks associated with solar radiation.

Solar radiation is a primary driver regulating urban climate and street energy balance. The various ecological and climatic issues resulting from rapid urbanization are widely impacting the sustainable development of human society (IPCC Climate Change, 2014). The rapid expansion of cities has led to the replacement of natural surfaces with artificial surfaces and buildings. The impervious surfaces in cities increase the absorption of solar radiation while reducing the cooling effect of natural vegetation transpiration, directly altering the heat balance and microclimate of urban areas, leading to the urban heat island effect (Oke, 1982). In the context of global warming, the urban heat island effect induced by urbanization is expected to intensify in the coming decades, bringing more frequent and severe heatwaves to urban residents (Ebi et al., 2021; Jay et al., 2021). Particularly in summer, the escalating heat island effect contributes to a general decline in thermal comfort for pedestrians on the streets. With the accelerated pace of urbanization and future climate warming, more urban residents are prone to heat stress. Therefore, better quantifying and assessing solar radiation at the urban street level will significantly enhance our understanding of the interactions between solar radiation, human health, and urban heat environment. This has important theoretical implications for China in formulating climate-adaptive urban planning.

On the other hand, especially in winter, high-density tall buildings significantly reduce the natural sunlight exposure on the surrounding streets. In Asia, sunlight visible radiation in street canyons under the shadow of tall buildings can decrease by up to 90% (Wai et al., 2015). Increasing evidence suggests that prolonged exposure to low levels of sunlight can have widespread health impacts, including vitamin D deficiency and reduced visual acuity (Boubekri, 2004). Therefore, urban high-density and dark street spaces not only have high energy consumption but are also unsustainable. Globally, around one billion people lack or have insufficient levels of vitamin D, leading to an increase in conditions related to weakened bones, such as rickets and osteomalacia (Naeem, 2010). Over 80% of our vitamin D comes from the skin exposed to sunlight, and the form of this vitamin obtained from sunlight lasts twice as long in the body as that absorbed through supplements, with no toxic risks. Since the 1960s, the global rise in myopia has been linked to insufficient sunlight. By 2050, half the world's population may be nearsighted. However, spending just 2 h outdoors in bright sunlight can help prevent myopia (Eppenberger & Sturm, 2020; Rose et al., 2008).

In addition, sunlight has various other benefits, such as its ability to kill bacteria and viruses, contributing to disease prevention in urban areas (Wimalawansa et al., 2019). Sunlit streets can create a brighter, more comfortable, and spacious atmosphere for residents, enhancing their quality of life and well-being (Gonçalves et al., 2019). Sunlit streets can also increase community vitality, promoting interaction among urban residents (Sciulli et al., 2023). Finally, sunlight stimulates plant growth, contributing to increased urban green coverage and environmental quality (Taniguchi et al., 2022). Therefore, accurately and efficiently quantifying urban street solar radiation and understanding its spatial and temporal distribution patterns and impact mechanisms are of crucial practical significance for evaluating residential quality and health risks related to street solar radiation. This understanding also contributes to advancing the construction of healthy and livable cities in China.

1.2.2 Street View Images in Urban Environmental Research

Street view images have become a crucial data source for urban analysis and geospatial data collection (Biljecki & Ito, 2021). These images not only reflect the real spatial and environmental perceptions from a human perspective but also provide a closer representation of real-life, capturing finer socioeconomic life details, thereby better reflecting the interaction between people and the environment in cities. Researchers can extract a wealth of urban information from street view data, such as urban spatial morphology and street composition elements, providing an indispensable source of novel urban data for street environment research. Street view images are considered a powerful complement to remote sensing images because they offer street-level three-dimensional spatial features that other data sources (such as aerial or satellite images) cannot provide (Zhang et al., 2018). These images contain rich information about urban infrastructure at the street level, allowing for an accurate reflection of the facade information of urban streets. With the introduction of various machine learning algorithms, accurate processing of street view images can be achieved, enabling effective identification of street elements such as sky, buildings, greenery, sidewalks, and lanes. In the last 5 years, street view images have shown extensive potential applications in urban analysis (Biljecki & Ito, 2021; Zhang et al., 2018). Specifically, their applications in the urban environment can be summarized in the following four aspects:

Urban Physical Environment

Street view images can reflect the real physical space of cities, including the spatial geometry of streets and street composition elements. For instance, Li et al. (2015) and Long and Liu (2017) quantified and analyzed the green view index of cities in the USA and China using street view images and image spectroscopy classification, respectively. Wang et al. (2018) measured the size of roadside trees on common city streets using Baidu street view images. Yao et al. (2019) proposed an algorithm to derive visual spatial relationships of various materials from continuous street view panoramas. Additionally, Gong et al. (2018) quantified the sky, building, and tree view factors of complex streets in Hong Kong's high-density urban areas using street view images and deep learning algorithms.

Urban Social Environment

By identifying urban elements in street view images and combining them with the socioeconomic environmental attributes mapped by these elements, assessments of the urban social environment can be conducted. For example, analyzing detailed information about vehicles in street view images can be used to estimate socioeconomic characteristics of 200 cities in the USA (Gebru et al., 2017). Law et al. (2020) assessed housing prices by training on street and satellite images. Innovative methods have been developed to automatically assess building damage and quality by combining street view images and deep learning models (Zhai & Peng, 2020).

Urban Perceptual Environment

Street view images can reflect human perceptual and behavioral attributes in the urban environment. For example, Naik et al. (2017) used Google street view images and computer vision methods to score changes in the appearance of streets in five major U.S. cities. Zhang et al. (2018) quantified individual perceptions of urban scenes using street view images and deep convolutional neural network models. Yao et al. (2019) introduced an innovative "human-machine adversarial" scoring method based on street view images. Furthermore, quantifying behavioral attributes can be achieved by assessing street view images perceived as more dangerous or more pleasant (Fan et al., 2023; He et al., 2017; Ye et al., 2019).

Urban Climate Environment

In the preliminary research conducted by the applicant (Gong et al., 2018; Gong, 2019; Gong et al., 2019), two key indicators influencing the urban microclimate environment in Hong Kong's high-density urban areas were quantified based on street view images. These indicators include the urban street sky, green, and building view factors as well as the spatial-temporal distribution of solar shortwave radiation energy in street canyons in the high-density areas of Hong Kong. The results were rigorously validated for high accuracy through on-site measurements and meteorological station data. Building on the foundation laid by the applicant's work, several scholars have also explored the application of street view images in quantifying parameters related to street solar radiation. For instance, tools have been developed based on street view images and 3D city models to calculate direct solar radiation (Liang et al., 2020). Studies have calculated street solar exposure lengths in the central urban area of Beijing using street view images (Du et al., 2020), investigated the variation of solar energy on streets in the central urban area of Chongqing (Deng et al., 2021), and estimated road photovoltaic capacity for supporting green transportation (Liu & Fei, 2021; Yang & Gong, 2025). However, existing studies evaluating solar-related parameters based on street view images assume ideal cloudless conditions or are focused on small-scale research within individual cities, limiting their applicability to the extensive assessment of real solar radiation on streets in China's high-density cities. The previous research conducted by the applicant in Hong Kong's high-density urban areas utilized a substantial amount of locally measured meteorological data for calculating street solar radiation. Consequently, the methods developed cannot be directly applied to high-density cities in different climatic zones across China. Therefore, the primary focus of this project is to develop a street canyon solar radiation estimation method based on street view images that is applicable to high-density cities in various climatic zones across China.

1.3 Research Questions

Throughout this book, the following five research questions will be mainly considered:

Research question 1: Is it technically possible to quantify the 3-D urban street canyon morphology effectively and accurately in a high-density urban environment for urban climate studies?

Google Street View (GSV) by Google company is a street-sensing technology that provides panoramic views along streets in the world. GSV freely serves millions of Google users with street-level panoramic imagery captured in hundreds of cities across four continents (Anguelov et al., 2010). Since GSV images directly capture urban streetscape and are available in many cities all over the world, this method provides a low cost and effective streetscape mapping approach for urban studies. There is recently proposed concept to use these publicly and freely

1.3 Research Questions

accessible street panoramic photographs from GSV to derive SVF of street canyons by projecting the panorama into fisheye images (Carrasco-Hernandez et al., 2015; Li et al., 2015; Liang et al., 2020). However, the previous study areas mainly focus on the cities where streetscape features are relatively simple with well-defined building and street structures. The feasibility and uncertainty of using GSV for estimating SVF and other street features, e.g. trees, building overhangs, and shade structures, in high-density complex urban form, are still not clear.

Research question 2: What is the accuracy and uncertainty of the widely used 3D-GIS model in quantifying the street canyon morphology?

3D-GIS Model simulation can produce spatially continuous VFs based on vectored buildings and rasterized digital 3-D surface models (Ratti & Richens, 2004; Gál et al., 2009; Chen et al., 2012). However, model data are difficult to accurately generate and are therefore not always available. The accuracy of VF estimations using model simulations depends heavily on the accuracy of the model in simulating the street environment. However, the street environment can be very complex, such as those in the high-density urban areas of Hong Kong. In particular, the street tree canopy, a major component of streetscapes, is hard to parameterize in models. It is possible to use the GSV-based view factors results, which are expected to be more accurate, as reference data to verify the 3D-model method results and optimize the results.

Research question 3: How to estimate the street-level solar radiation effectively and accurately at large scale for urban climate studies?

In a high-density urban environment, solar radiation goes into street canyons through their sky opening, while buildings and trees are the two main obstructions of the solar light path that prevents direct sunlight from reaching the ground at particular times of the day. Since GSV images provide a direct mapping of street morphologies, they can be used to quantify the sky opening and the obstructions by building and trees. Moreover, the high-density urban street morphologies, which are usually complex and have large spatial variations, are not well captured by models but can be fully characterized by GSV images. Therefore, extending from our *Research question 1*, this study further proposes using the street geometries characterized using GSV images to quantify the street-level solar irradiance.

Research question 4: How does the urban canyon geometry affect the street-level solar radiation?

Street canyons geometry, including street orientation and SVF, has an important impact on the street-level solar radiation. The SVF is the fraction of sky in the upper hemisphere of a street canyon and determines the total incoming solar radiation that reaches the top of a street canyon. The orientation of a street canyon affects the timing of exposure to direct sunlight, particularly in a high-density environment with high H/W ratio. In a North-South street canyon, the street surface will be exposed to direct sunlight near mid-day but shaded in other times. However, in an East–West street canyon, it is more likely to be exposed all day from the morning to the afternoon in low latitude cities like Hong Kong (Erell et al., 2014). Therefore, the horizon obstructions of street canyons, including buildings and trees, should be described when calculating the solar radiation at street level.

Research question 5: What are the spatial and temporal patterns of street canyon morphologies and street-level solar radiation in high-density urban areas of Hong Kong? How may these patterns help improve urban planning and design at street level in a high-density urban environment?

Characterization of urban street morphology and understanding of street-level energy balance are important in aiding the urban planning and design process at street level by policymakers. In this study, the street morphology in the high-density urban area of Hong Kong can be characterized using street view factors, while the street-level energy balance can be driven by the solar energy input which is quantified by the street-level solar radiation. By analyzing the maps of VFs and solar radiation and understanding their spatial and temporal patterns, we can identify street-level hotspots with unsatisfactory urban morphological setting or solar radiation exposure. These hotspots may draw attention from the policymakers to understand directly the provision of the realistic conditions and problems by visualization. These results and analysis can provide the quantification indexes of urban planning and urban climate studies, which can further help make the planning indexes for developing a better urban environment.

1.4 Research Objectives

View factors for the sky, trees, and buildings are three important parameters of the urban outdoor environment that describe the geometrical relation between different surfaces from the perspective of radiative energy transfer. The developed approach for analyzing sky, tree, and building view factors in a 3-D street environment will play an important role in relating science-based evidence for urban climatic studies and decision-making in urban planning and design processes. The maps of street-level solar radiation may help researchers investigate the interactions between solar radiation, human health, and urban thermal balance in high-density urban environments.

The present book aims to develop an efficient, high-accurate, and open-access approach to extract 3-D urban morphology information, and then retrieve and validate typical urban morphological parameters for urban climatic applications; and reconstruct urban canyon irradiance, to create more realistic descriptions of the street-level radiative environment, which may find the useful applications in urban climatic and human exposure studies. The study will use the high-density urban areas of Hong Kong, i.e., Kowloon area and Hong Kong Island, as the study area. Hong Kong, located in monsoon Asia, has a high-rise, high-density, and compact urban morphology with high building-height-to-street-width (H/W) ratio. The specific objectives of this research are as follows:

Research objective 1: To develop an effective and accurate approach to estimate street view factors for sky openness, tree canopy, and building in a high-density urban environment.

This is to develop an approach for accurately estimating sky view factor (SVF), tree view factor (TVF), and building view factor (BVF) of street canyons in the

1.4 Research Objectives

high-density urban environment of Hong Kong using publicly available Google Street View (GSV) images and a deep learning algorithm for extraction of street features (sky, trees, and buildings).

Research objective 2: To verify the developed view factor estimation method using field measurements and investigate the uncertainty of 3D-GIS model by comparing to the GSV-based method.

Verification using reference data of hemispheric photography from field surveys in compact high-rise and low-rise areas is conducted. This is the first reported use of hemispheric photography for direct verification in a GSV-based streetscape study. Furthermore, a comparison between GSV-based and 3D-GIS-based SVFs investigate the uncertainty of 3D-GIS model by comparing to a GSV-based method and to explain the success and failure of 3D-GIS model in characterizing streetscape in a high-density environment.

Research objective 3: To further develop an effective and accurate approach to estimate the solar irradiance of street canyons, including its direct and diffuse components.

This is to develop a method for calculating the street-level solar radiation using Google Street View (GSV) images and investigates the spatiotemporal patterns of street-level solar global, direct, and diffuse radiation in a high-density urban environment. GSV images are used to characterize the street morphology from which diurnal solar path and solar radiation exposure can be estimated in a street canyon. Verifications of our developed method using full-sky observatory and field measurements in a high-density street canyon show that both the calculated clear-sky and all-sky street-level solar radiation well capture its diurnal and seasonal cycle.

Research objective 4: To investigate the impact of urban canyon morphology and street geometry on the street-level solar radiation and its spatiotemporal pattern in a high-density urban environment.

The spatial variability of street-level solar radiation is closely related to SVF. With larger SVF, the sky opening will be wider and therefore more solar radiation is coming into the streets. Furthermore, the street orientation has a significant impact on the solar radiation received in a high-density street canyon. In general, street canyons with West-East orientation receives higher solar radiation during summer and, however, lower during winter compared to those with South-North orientation. With the generated maps the street-level solar irradiance, including direct and diffuse components, we can quantify the impact of urban canyon morphology and street geometry on the street-level solar radiation in the high-density urban areas of Hong Kong.

Research objective 5: To identify street-level hotspots with unsatisfactory urban morphological setting or solar radiation exposure according to the above results and provide suggestions for better urban planning and design at street level in a high-density urban environment.

The resulted maps of street view factors and street-level solar irradiance will provide useful datasets for studying the interactions between solar radiation, human health and the urban thermal balance in high-density urban environments. Easy access to street-level greenery, wide sky opening, and adequate exposure to solar radiation have been shown to be important for promoting physical activities and improving public health (Hunter et al., 2015; Schipperijn et al., 2013). With the spatial and temporal mappings of street morphology and street-level solar radiation, street-level hotspots can be identified with insufficient street-level greenery, very small sky opening, excessive or insufficient solar exposure. Policymakers can then be informed of what needs to be modified in the practical planning and design stage. Policymakers can also consequently identify appropriate planning strategies based on particular requirements and the extent of modification required in particular planning or design cases to mitigate negative effects on the surrounding environments. This research provides a scientific understanding of decision-making during the urban planning and design practices.

1.5 Book Structure

The core of the book is divided into two parts. The first part is devoted to the view factors quantification using Google Street View (GSV), including the sky, tree, and building view factors at street level. The second part is devoted to solar radiation estimation based on GSV, including global, direct, and diffuse radiation at street level. This book is organized with the following seven chapters:

- Chapter 1 is the introduction of this book, including the research background, research questions, and research objectives.
- Chapter 2 presents the research background and the theoretical framework of this study.
- Chapter 3 presents the study area, used data, and the methodologies of this study. The street-sensing method includes two major parts: Part I introduces the method to estimate urban 3-D morphology, including sky openness, tree canopy, and building density using GSV images; and Part II further introduces the GSV-based method to estimate solar radiation, including global, direct, and diffuse radiation.
- Chapter 4 presents the results of verifying and mapping the street view factors (VFs) of street canyon using GSV images, including sky view factor (SVF), building view factor (BVF), and tree view factor (TVF), and further analysis the spatial patterns of street view factors in high-density areas of Hong Kong
- Chapter 5 presents the results of verifying and quantifying the solar radiation of street canyons using GSV images, including global solar radiation, and its direct and diffuse components, and further analyses the spatiotemporal patterns of solar radiation of street canyons in high-density areas of Hong Kong.
- Chapter 6 presents the implementation based on the results of this study in urban planning and design practices for researchers, policymaker and urban planners.
- Chapter 7 provides a summary of contributions, comments on the strengths and limitations of this developed approach and outlines the prospects for future works.

References

Akbari, H., Pomerantz, M., & Taha, H. (2001). Cool surfaces and shade trees to reduce energy use and improve air quality in urban areas. *Solar Energy, 70*(3), 295–310. https://doi.org/10.1016/S0038-092X(00)00089-X

Anguelov, D., Dulong, C., Filip, D., Frueh, C., Lafon, S., Lyon, R., ... Weaver, J. (2010). Google street view: Capturing the world at street level. *Computer, 43*(6), 32–38. https://doi.org/10.1109/MC.2010.170

Biljecki, F., & Ito, K. (2021). Street view imagery in urban analytics and GIS: A review. *Landscape and Urban Planning, 215*, 104217. https://doi.org/10.1016/j.landurbplan.2021.104217

Boubekri, M. (2004). An argument for daylighting legislation because of health. *Journal of the Human-Environment System, 7*(2), 51–56. https://doi.org/10.1618/jhes.7.51

Brager, G. S., & de Dear, R. J. (1998). Thermal adaptation in the built environment: A literature review. *Energy and Buildings, 27*(1), 83–96. https://doi.org/10.1016/S0378-7788(97)00053-4

Carrasco-Hernandez, R., Smedley, A. R. D., & Webb, A. R. (2015). Using urban canyon geometries obtained from Google street view for atmospheric studies: Potential applications in the calculation of street level total shortwave irradiances. *Energy and Buildings, 86*(Supplement C), 340–348. https://doi.org/10.1016/j.enbuild.2014.10.001

Chen, L., Ng, E., An, X., Ren, C., Lee, M., Wang, U., & He, Z. (2012). Sky view factor analysis of street canyons and its implications for daytime intra-urban air temperature differentials in high-rise, high-density urban areas of Hong Kong: A GIS-based simulation approach. *International Journal of Climatology, 32*(1), 121–136. https://doi.org/10.1002/joc.2243

Deng, M., Yang, W., Chen, C., Wu, Z., Liu, Y., & Xiang, C. (2021). Street-level solar radiation mapping and patterns profiling using Baidu street view images. *Sustainable Cities and Society, 75*, 103289. https://doi.org/10.1016/j.scs.2021.103289

Du, K., Ning, J., & Yan, L. (2020). How long is the sun duration in a street canyon? Analysis of the view factors of street canyons. *Building and Environment, 172*, 106680. https://doi.org/10.1016/j.buildenv.2020.106680

Ebi, K. L., Capon, A., Berry, P., Broderick, C., de Dear, R., Havenith, G., & Jay, O. (2021). Hot weather and heat extremes: Health risks. *The Lancet, 398*(10301), 698–708. https://doi.org/10.1016/S0140-6736(21)01208-3

Eppenberger, L. S., & Sturm, V. (2020). The role of time exposed to outdoor light for myopia prevalence and progression: A literature review. *Clinical Ophthalmology, 14*, 1875–1890. https://doi.org/10.2147/OPTH.S245192

Erell, E., Pearlmutter, D., Boneh, D., & Kutiel, P. B. (2014). Effect of high-albedo materials on pedestrian heat stress in urban street canyons. *Urban Climate, 10*, 367–386. https://doi.org/10.1016/j.uclim.2013.10.005

Fan, Z., Zhang, F., Loo, B. P., & Ratti, C. (2023). Urban visual intelligence: Uncovering hidden city profiles with street view images. *Proceedings of the National Academy of Sciences, 120*(27), e2220417120. https://doi.org/10.1073/pnas.2220417120

Farrar, M. D., Webb, A. R., Kift, R., Durkin, M. T., Allan, D., Herbert, A., & Rhodes, L. E. (2013). Efficacy of a dose range of simulated sunlight exposures in raising vitamin D status in south Asian adults: Implications for targeted guidance on sun exposure. *The American Journal of Clinical Nutrition, 97*(6), 1210–1216. https://doi.org/10.3945/ajcn.112.052639

Gál, T., Lindberg, F., & Unger, J. (2009). Computing continuous sky view factors using 3D urban raster and vector databases: Comparison and application to urban climate. *Theoretical and Applied Climatology, 95*, 111–123. https://doi.org/10.1007/s00704-007-0362-9

Gebru, T., Krause, J., Wang, Y., Chen, D., Deng, J., Aiden, E. L., & Fei-Fei, L. (2017). Using deep learning and Google street view to estimate the demographic makeup of neighborhoods across the United States. *Proceedings of the National Academy of Sciences, 114*(50), 13108–13113. https://doi.org/10.1073/pnas.17000351

Gonçalves, G., Sousa, A., Sousa, C., Jesus, F., & Afonso, E. (2019). Effects of sunlight on psychological Well-being, job satisfaction and confinement perception of workplace: The case of shopkeepers and marketers. In *Occupational and environmental safety and health* (pp. 573–580). Springer.

Gong, F.-Y. (2019). *Mapping street canyon morphology and solar radiation in high-density urban environments using street sensing approach.* The Chinese University of Hong Kong.

Gong, F.-Y., Zeng, Z.-C., Ng, E., & Norford, L. K. (2019). Spatiotemporal patterns of street-level solar radiation estimated using Google street view in a high-density urban environment. *Building and Environment, 148,* 547–566. https://doi.org/10.1016/j.buildenv.2018.10.025

Gong, F.-Y., Zeng, Z.-C., Zhang, F., Li, X., Ng, E., & Norford, L. K. (2018). Mapping sky, tree, and building view factors of street canyons in a high-density urban environment. *Building and Environment, 134,* 155–167. https://doi.org/10.1016/j.buildenv.2018.02.042

He, L., Páez, A., & Liu, D. (2017). Built environment and violent crime: An environmental audit approach using Google street view. *Computers, Environment and Urban Systems, 66,* 83–95. https://doi.org/10.1016/j.compenvurbsys.2017.07.004

Hunter, R. F., Christian, H., Veitch, J., Astell-Burt, T., Hipp, J. A., & Schipperijn, J. (2015). The impact of interventions to promote physical activity in urban green space: A systematic review and recommendations for future research. *Social Science & Medicine, 124,* 246–256. https://doi.org/10.1016/j.socscimed.2014.11.051

Institute of Population and Labor Economics, Chinese Academy of Social Sciences. (2021). *Population and labor green book: Report on China's population and labor issues.* Social Sciences Academic Press.

IPCC Climate Change. (2014). *Synthesis report. Contribution of working groups I, II and III to the fifth assessment report of the intergovernmental panel on climate change.* IPCC.

Jakubiec, J. A., & Reinhart, C. F. (2013). A method for predicting city-wide electricity gains from photovoltaic panels based on LiDAR and GIS data combined with hourly Daysim simulations. *Solar Energy, 93,* 127–143. https://doi.org/10.1016/j.solener.2013.03.022

Jay, O., Capon, A., Berry, P., Broderick, C., de Dear, R., Havenith, G., & Ebi, K. L. (2021). Reducing the health effects of hot weather and heat extremes: From personal cooling strategies to green cities. *The Lancet, 398*(10301), 709–724. https://doi.org/10.1016/S0140-6736(21)01209-5

Johansson, E. (2006). Influence of urban geometry on outdoor thermal comfort in a hot dry climate: A study in fez, Morocco. *Building and Environment, 41*(10), 1326–1338. https://doi.org/10.1016/j.buildenv.2005.05.022

Law, S., Seresinhe, C. I., Shen, Y., & Gutierrez-Roig, M. (2020). Street-frontage-net: Urban image classification using deep convolutional neural networks. *International Journal of Geographical Information Science, 34*(4), 681–707. https://doi.org/10.1080/13658816.2019.1627634

Li, X., Zhang, C., Li, W., Ricard, R., Meng, Q., & Zhang, W. (2015). Assessing street-level urban greenery using Google Street View and a modified green view index. *Urban Forestry & Urban Greening, 14*(3), 675–685. https://doi.org/10.1016/j.ufug.2015.06.006

Liang, J., Gong, J., Xie, X., & Sun, J. (2020). Solar3D: An open-source tool for estimating solar radiation in urban environments. *ISPRS International Journal of Geo-Information, 9*(9), 524. https://doi.org/10.3390/ijgi9090524

Liu, Z., & Fei, T. (2021). Road PV production estimation at city scale: A predictive model towards feasible assessing regional energy generation from solar roads. *Journal of Cleaner Production, 321,* 129010. https://doi.org/10.1016/j.jclepro.2021.129010

Long, Y., & Liu, L. (2017). How green are the streets? An analysis for central areas of Chinese cities using Tencent street view. *PLoS One, 12*(2), e0171110. https://doi.org/10.1371/journal.pone.0171110

Naeem, Z. (2010). Vitamin D deficiency-an ignored epidemic. *International Journal of Health Sciences, 4*(1), 5. https://doi.org/10.1093/jn/135.11.2739S

Naik, N., Kominers, S. D., Raskar, R., Glaeser, E. L., & Hidalgo, C. A. (2017). Computer vision uncovers predictors of physical urban change. *Proceedings of the National Academy of Sciences, 114*(29), 7571–7576. https://doi.org/10.1073/pnas.1614069114

References

Oke, T. R. (1982). The energetic basis of the urban heat Island. *Quarterly Journal of the Royal Meteorological Society, 108*(455), 1–24. https://doi.org/10.1002/qj.49710845502

Oke, T. R. (1988). The urban energy balance. *Progress in Physical Geography: Earth and Environment, 12*(4), 471–508. https://doi.org/10.1177/030913338801200401

Oleson, K. (2011). Contrasts between urban and rural climate in CCSM4 CMIP5 climate change scenarios. *Journal of Climate, 25*(5), 1390–1412. https://doi.org/10.1175/JCLI-D-11-00098.1

Ratti, C., & Richens, P. (2004). Raster analysis of urban form. *Environment and Planning B: Planning and Design, 31*(2), 297–309. https://doi.org/10.1068/b2665

Rhodes, L. E., Webb, A. R., Fraser, H. I., Kift, R., Durkin, M. T., Allan, D., & Berry, J. L. (2010). Recommended summer sunlight exposure levels can produce sufficient ($\geq$20ngml−1) but not the proposed optimal ($\geq$32ngml−1) 25(OH)D levels at UK latitudes. *Journal of Investigative Dermatology, 130*(5), 1411–1418. https://doi.org/10.1038/jid.2009.417

Rose, K. A., Morgan, I. G., Ip, J., Kifley, A., Huynh, S., Smith, W., & Mitchell, P. (2008). Outdoor activity reduces the prevalence of myopia in children. *Ophthalmology, 115*(8), 1279–1285. https://doi.org/10.1016/j.ophtha.2007.12.019

Saito, I., Ishihara, O., & Katayama, T. (1990). Study of the effect of green areas on the thermal environment in an urban area. *Energy and Buildings, 15*(3), 493–498. https://doi.org/10.1016/0378-7788(90)90026-F

Sanusi, R., Johnstone, D., May, P., & Livesley, S. J. (2016). Street orientation and side of the street greatly influence the microclimatic benefits street trees can provide in summer. *Journal of Environmental Quality, 45*(1), 167–174. https://doi.org/10.2134/jeq2015.01.0039

Schipperijn, J., Bentsen, P., Troelsen, J., Toftager, M., & Stigsdotter, U. K. (2013). Associations between physical activity and characteristics of urban green space. *Urban Forestry & Urban Greening, 12*(1), 109–116. https://doi.org/10.1016/j.ufug.2012.12.002

Sciulli, N., Goullet, D., & Snell, T. (2023). Working from home with a view of nature (and sunlight) benefits people's Well-being. *Ecopsychology, 15*(1), 69–80. https://doi.org/10.1089/eco.2022.003

Taniguchi, K., Takano, M., Tobari, Y., Hayano, M., Nakajima, S., Mimura, M., & Noda, Y. (2022). Influence of external natural environment including sunshine exposure on public mental health: A systematic review. *Psychiatry International, 3*(1), 91–113. https://doi.org/10.3390/psychiatryint3010008

Thorsson, S., Lindberg, F., Björklund, J., Holmer, B., & Rayner, D. (2010). Potential changes in outdoor thermal comfort conditions in Gothenburg, Sweden due to climate change: The influence of urban geometry. *International Journal of Climatology, 31*(2), 324–335. https://doi.org/10.1002/joc.2231

UNFPA. (2017). *Urbanization*. Retrieved October 25, 2018.

Wai, K. M., Yu, P. K., & Lam, K. S. (2015). Reduction of solar UV radiation due to urban high-rise buildings–a coupled modelling study. *PLoS One, 10*(8), e0135562. https://doi.org/10.1371/journal.pone.0135562

Wang, W., Xiao, L., Zhang, J., Yang, Y., Tian, P., Wang, H., & He, X. (2018). Potential of internet street-view images for measuring tree sizes in roadside forests. *Urban Forestry & Urban Greening, 35*, 211–220. https://doi.org/10.1016/j.ufug.2018.08.011

Webb, A. R. (2006). Considerations for lighting in the built environment: Non-visual effects of light. *Energy and Buildings, 38*(7), 721–727. https://doi.org/10.1016/j.enbuild.2006.03.004

Wimalawansa, S. J. (2019). Vitamin D deficiency: Effects on oxidative stress, epigenetics, gene regulation, and aging. *Biology, 8*(2), 30. https://doi.org/10.3390/biology8020030

Yang, Z., & Gong, F.-Y. (2025). Utilizing street view images to estimate solar energy potential for photovoltaic-powered buses. *Applied Geography, 177*, 103567. https://doi.org/10.1016/j.apgeog.2025.103567

Yao, Y., Liang, Z., Yuan, Z., Liu, P., Bie, Y., Zhang, J., & Guan, Q. (2019). A human-machine adversarial scoring framework for urban perception assessment using street-view images. *International Journal of Geographical Information Science, 33*(12), 2363–2384. https://doi.org/10.1080/13658816.2019.1669872

Ye, Y., Richards, D., Lu, Y., Song, X., Zhuang, Y., Zeng, W., & Zhong, T. (2019). Measuring daily accessed street greenery: A human-scale approach for informing better urban planning practices. *Landscape and Urban Planning, 191*, 103434. https://doi.org/10.1016/j.landurbplan.2019.103434

Zhai, W., & Peng, Z. R. (2020). Damage assessment using Google street view: Evidence from hurricane Michael in Mexico beach, Florida. *Applied Geography, 123*, 102252. https://doi.org/10.1016/j.apgeog.2020.102252

Zhang, F., Zhou, B., Liu, L., Liu, Y., Fung, H. H., Lin, H., & Ratti, C. (2018). Measuring human perceptions of a large-scale urban region using machine learning. *Landscape and Urban Planning, 180*, 148–160. https://doi.org/10.1016/j.landurbplan.2018.08.020

Zhao, L., Oleson, K., Bou-Zeid, E., Krayenhoff, E. S., Bray, A., Zhu, Q., & Oppenheimer, M. (2021). Global multi-model projections of local urban climates. *Nature Climate Change, 11*(2), 152–157. https://doi.org/10.1038/s41558-020-00958-8

Zielinska-Dabkowska, K. M., & Xavia, K. (2019). Protect our right to light. *Nature, 568*(7753), 451–453. https://www.nature.com/articles/d41586-019-01238-y

Chapter 2
Urban Canyon Morphology and Solar Radiation Dynamics

Contents

2.1	Overview	16
2.2	Street Canyon Morphology and Quantification	16
	2.2.1 Street View Factors as Morphology Indicators in Urban Climate Study	16
	2.2.2 Methods of Quantifying Street View Factors	18
2.3	Street Canyon Solar Radiation and Estimations	19
	2.3.1 Physical Basis of Street-Level Solar Irradiance	20
	2.3.2 Methods of Estimating Solar Radiation	23
2.4	Summary	24
References		25

Abstract Urban street canyons significantly influence local climate conditions by regulating solar radiation and thermal energy exchange. This chapter explores the geometric characteristics of urban street canyons and their role in modifying solar radiation exposure at the street level. Key view factors, including the sky view factor (SVF), tree view factor (TVF), and building view factor (BVF), serve as primary indicators of urban morphology and are critical for understanding urban thermal environments. This chapter reviews different methods for quantifying these view factors, including 3D model simulations, fisheye photographic analysis, and street-level sensing through panoramic images such as Google Street View. Additionally, this chapter delves into the physical basis of solar radiation at the street level, discussing direct, diffuse, and reflected radiation within high-density urban environments. The integration of these view factors into solar radiation estimation provides insights into the impact of urban geometry on microclimate conditions, urban heat island effects, and outdoor thermal comfort. The findings contribute to urban planning strategies aimed at improving climate resilience and human well-being in dense metropolitan areas.

Keywords Urban street canyon · Solar radiation · Sky view factor (SVF) · Google street view (GSV) · Urban microclimate

2.1 Overview

This chapter describes the basic concepts of urban canyon morphology related to climate study and solar radiation theory that will be used in this book. Meanwhile, this chapter reviews recent methods to quantify the urban morphology and solar irradiation at street level. The physical basis of the interaction between solar radiation and the atmosphere and the calculation of street-level solar radiation and its key components incident on a street surface is also introduced in this chapter.

Throughout this book, the terminology term *"urban street canyon"* is any location in a street that is blocked by obstructions, including building façades, bridge construction, or vegetations. The urban street canyon is also a geometric abstraction that illustrates how the basic street geometries regulate access to solar radiation (Strømann-Andersen & Sattrup, 2011); we consistently use *"solar irradiance"* when referring to the intensity of solar radiation (in unit of W/m^2) received by a horizontal surface; and we use *"solar irradiation"* when referring to the integration of solar irradiance over a certain time range (in unit of MJ/m^2). We also use intensity units, W/m^2, when referring to solar radiation independent of its incidence on a given surface.

2.2 Street Canyon Morphology and Quantification

Street view factors (VFs) for the sky, trees, and buildings are three important parameters of urban outdoor environments. They describe the geometrical relation between different urban street components from perspectives of radiative energy transfer which plays a key role in urban thermal environments. Sky view factor (SVF), tree view factor (TVF), and building view factor (BVF) are defined as the geometric ratio of the amount of the sky, trees, and buildings seen, respectively, from a given surface point to the overlying hemisphere subtended by a horizontal surface (Johansson, 2006; Oke, 1987). A thorough quantification and understanding of the physical streetscape using view factors, including its features and dynamics, would offer great utility to urban planners and climatologists investigating the urban environment, its physical and social interactions, and implications for human well-being.

2.2.1 Street View Factors as Morphology Indicators in Urban Climate Study

Sky View Factor (SVF)

Sky view factor (SVF), as a geometric quantification of the degree of sky visibility within street canyons, is a commonly used urban geometry indicator in morphology-oriented urban climate studies. As an effective indicator of nocturnal urban radiation balance, SVF characterizes the ratio of received (or emitted) radiation by an urban street to the total radiation emitted (or received) by the entire hemispheric radiation environment (Watson & Johnson, 1987). Therefore, SVF is an important geometrical parameter for the studies of urban microclimate (Bourbia & Boucheriba, 2010; Johansson, 2006), nocturnal urban heat island (UHI) effect (Oke, 1981; Unger, 2004), urban thermal comfort (Johansson, 2006; Krüger et al., 2011), and urban air pollution (Tang & Wang, 2007). Urban morphology has been shown in many cities to be critical in mitigating nocturnal UHI than daytime UHI. This is due to the influence of the slower wind in the night time (Morris et al., 2001). The emission of long-wave radiation is largely determined by the surface-temperature distribution, and this is known to be strongly affected by tree cover and vegetation in general (Oke, 1989). At night, the energy of the outgoing net radiation from a vegetation surface is fed from the thermal heat flux and the latent heat flux. Therefore, the temperature around the vegetation area is lower than that around the built-up areas (Wong & Yu, 2005).

Tree View Factor (TVF)

Street tree canopy, quantified by TVF, has instrumental ecological service functions such as UHI mitigation due to its contribution to reduce urban temperature (Li & Norford, 2016; Ng et al., 2012; Oliveira et al., 2011; Shashua-Bar et al., 2010). The trees' cooling effect come from tree shading, which reduces the radiation reaching ground level (Sutherland & Bartholic, 1977; Parisi et al., 2000), and evaporative cooling from leaf surfaces (Shashua-Bar & Hoffman, 2002). In addition, urban street trees have been found to absorb airborne pollutions and therefore decrease road traffic emissions (Abhijith & Gokhale, 2015) and improve the walkability of streets (Klemm et al., 2015). Therefore, the proportion of street tree cover can be used to evaluate the benefits from ecosystem service provisions at different areas of a city (Richards & Edwards, 2017).

Building View Factor (BVF)

Many urban materials for buildings have a relatively high heat capacity and surface thermal admittance, which make them efficiently accept and retain heat during daytime and release it at night leading to a strong UHI effect (Fernando, 2012; Roth & Chow, 2012). This study uses BVF to quantify the impact of buildings in the urban radiative balance. These three VFs interact with each other in balancing urban radiation. SVF is a combination factor of buildings and trees in influencing the air temperature (Wong & Yu, 2005). Street trees reduce SVF by providing shading to the environment, resulting in the reduction of nighttime net long-wave loss (Oke, 1989). Buildings emit a greater amount of long-wave radiation compared to the cool sky and trees (Henderson-Sellers, 1995). Therefore, urban street canyons with a higher BVF will yield a larger net long-wave radiation.

2.2.2 Methods of Quantifying Street View Factors

Methods for estimating view factors or parameterizing urban geometries of street canyons can be grouped into the following three types: i.e., (1) digital 3-D surface models, including digital elevation models (DEMs) or 3D GIS model; (2) fisheye photographs, and (3) street-sensing method based on street view panoramas. The advantages and disadvantages of these methods are described as follows:

3-D Model Simulation

This method can produce spatially continuous SVF based on vectored buildings and rasterized digital 3-D surface models, such as 3-D GIS-based and DSM-based models (Chen et al., 2012; Gál et al., 2009; Ratti & Richens, 2004). The 3-D GIS-based model incorporates digital elevation vector layers of the urban surface and the 3-D building structure above the surface. DSM-based model of cities incorporates digital raster maps in which each pixel characterizes the height of natural or/and built features on the surface. Therefore, 3-D GIS-based and DSM-based models support rapid approximation calculation of street features for the large-area urban environment.

However, there are two main disadvantages. Firstly, model data are difficult to accurately generate and are therefore not always available, especially for cities due to the high costs of acquiring building height information. Secondly, the accuracy of street view factor estimations using model simulations depends heavily on the accuracy of the model in simulating the street environment. In 3-D GIS model, the information of street tree canopy is not included or well characterized; while in DSM model the averaged building height in each pixel may lead to large inaccuracy if the resolution is low. Also, the different shapes of street tree canopy will be hard to be parameterized in models. Therefore, a very high resolution is needed to characterize the street-level environment, which is expensive and not realistic for worldwide cities. In addition, the street environment can be very complex, such as those in the high-density urban areas of Hong Kong. In particular, the street tree canopy, a major component of streetscapes, is hard to parameterize in models.

Photographic Method

This method uses a fisheye lens to take on-site photographs which project the hemispheric environment onto a circular plane. Different street features are then extracted from the fisheye image to calculate the VFs. This method provides a direct and accurate measurement of SVF (Anderson, 1964; Steyn, 1980). However, taking on-site fisheye images to represent cities at large scale usually requires fieldwork that is time- and effort-consuming. Therefore, this method is suitable only for small-scale study; Fortunately, the Google Street View image database provides a means for accessing street-level imagery. The methods of projecting the Google panorama images from cylindrical to azimuthal projection to generate the fisheye images is described in Sect. 3.3.1.

Street-Sensing Method Based on Street View Panoramas

This recently proposed concept uses publicly and freely accessible street panoramic photographs, e.g., Google Street View (GSV) images, to derive sky view factor (SVF) of street canyons by projecting the panorama into fisheye images (Carrasco-Hernandez et al., 2015; Gong et al., 2018; Li et al., 2018; Liang et al., 2017; Middel et al., 2017). Since GSV images directly capture urban streetscape and are available in many cities all over the world, this method provides a low cost and effective streetscape mapping approach for urban studies.

Carrasco-Hernandez et al. (2015) proposed using the GSV images to calculate the street-level total shortwave irradiances estimated from an open-source panorama generating tool. Li et al. (2018) showed two examples demonstrating the usage of GSV images and 3-D building data in mapping street trees based on RGB method, while Liang et al. (2017) provides a proof-of-concept study to show the reliability of using street panorama images in estimating SVF. Middel et al. (2017) used GSV estimated SVF to validate the synthetic hemispherical fisheye photos generated from a developed web-based tool using Google Earth 3-D data for urban areas.

However, the previous study areas mainly focus on the cities where streetscape features are relatively simple (compared to high-density urban areas of Hong Kong in this study) with well-defined building and street structures. The feasibility and uncertainty of using Google Street View for estimating SVF in such high-density context are still not clear; Second, these studies lack independent verifications of their calculations. Third, other street view factors, including 3-D tree canopy and buildings have not been effectively and accurately quantified using street view images. Gong et al. (2018) developed an approach to map all street view factors including sky, building, and tree view factors using pyramid scene parsing network (PSPNet) and demonstrated the high accuracy and effectiveness of this method by direct comparison with hemispheric photography based on the approach proposed in this book.

2.3 Street Canyon Solar Radiation and Estimations

Solar radiation is the main driver in regulating urban climate and street-level thermal energy balance (Oke, 1988b). Solar radiation in urban areas has been extensively investigated in different fields including urban meteorology (Oleson, 2011), photovoltaic generation (Jakubiec & Reinhart, 2013; Kardakos et al., 2013), urban heat island effect (Oke, 1982) and its related issues such as thermal comfort, human health issues due to UV exposure (Farrar et al., 2013; Rhodes et al., 2010; Webb,

2006), and urban microclimate study (Sanusi et al., 2016). Due to the increasing trend in urbanization and an expected warmer climate in the near future, more and more residents are prone to heat stress in cities (Akbari et al., 2001; Thorsson et al., 2010). Therefore, better quantification of solar radiation at urban street levels will greatly improve our understandings of the interactions between solar radiation, human health, and the urban thermal environment.

2.3.1 Physical Basis of Street-Level Solar Irradiance

There are three key components in the solar radiation incident on a surface at street level in a high-density environment as shown in Fig. 2.1: the direct radiation, the radiation diffused by atmospheric molecules, aerosols or clouds, and radiation reflected from buildings and ground (Matzarakis et al., 2010; Oke, 1988a, 1988b). As shown as beam **A,** the direct radiation is rays that come to the bottom of the street canyon in a straight line from the direction of the sun. Diffuse irradiance is the amount of radiation received by the bottom of the street canyon that has been scattered by atmospheric molecules shown as B_1, particles shown as B_2, and clouds shown as B_3 in the atmosphere and potentially comes from all directions. The reflected radiation is the sunlight that has been reflected off of non-atmospheric surfaces such as the ground (shown as C_1) and buildings (shown as C_2). Multiple scatterings and reflections by urban materials and atmosphere within the urban street canyons are not shown here. As indicated in Fig. 2.2, three key factors influence the street-level solar irradiance in a high-density environment:

- **Solar geometries**: including the solar zenith angle (SZA) and solar azimuth angle (SAA). In Hong Kong (22°17′7.87″N, 114°9′27.68″E), the SZA at local noon time changes from 45.5° in the winter and 3.5° in the summer. The solar geometries determine the associated solar irradiance on a horizon surface due to seasonal variation;

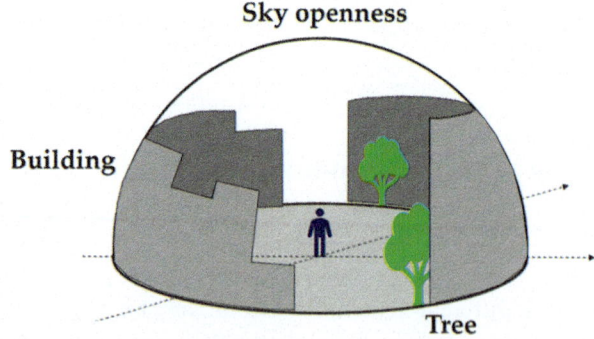

Fig. 2.1 The physical streetscape using view factors, including the main street components, sky, trees, buildings and ground in a high-density environment

2.3 Street Canyon Solar Radiation and Estimations

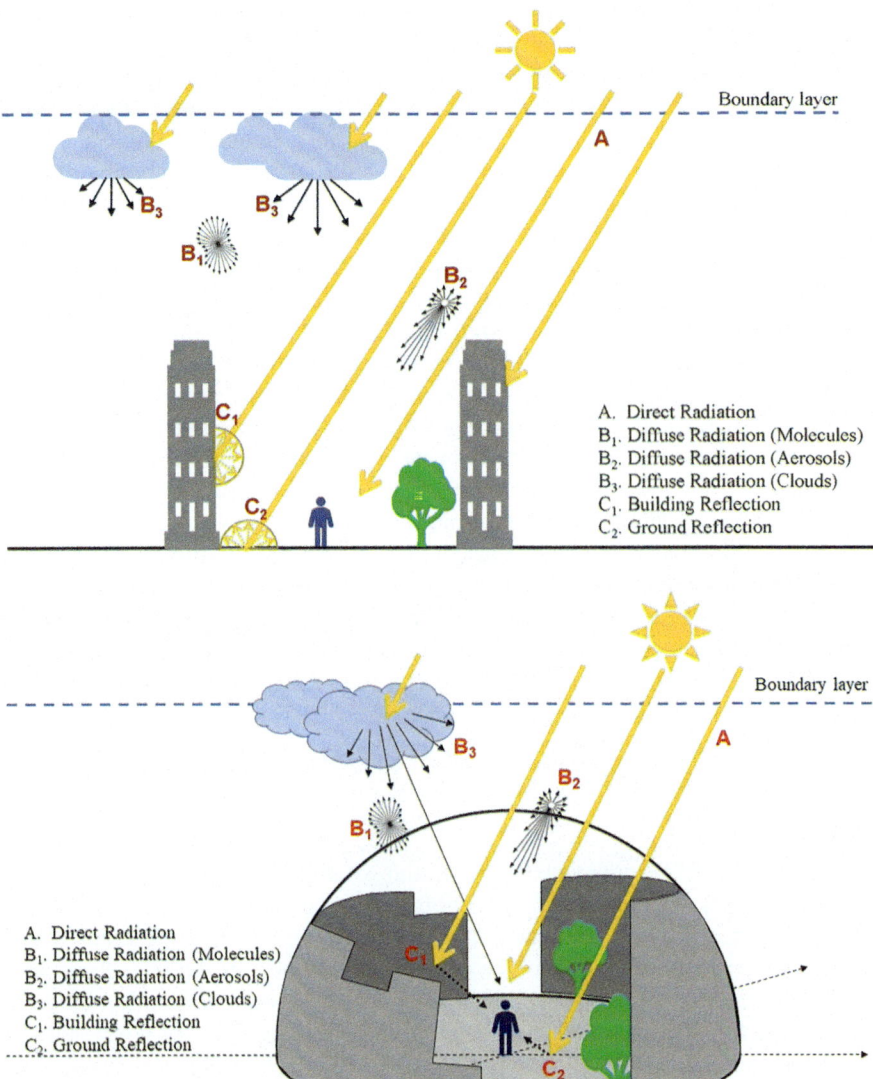

Fig. 2.2 Physical basis of incident solar radiation at street level in a high-density environment. As indicated, the street-level solar radiation incident on a surface includes direct and diffuse components from atmospheric molecules (Rayleigh scattering) and aerosols (Mie scattering), and clouds (Liou, 2002; Ross, 1981), and the reflected radiation from buildings and ground. The radiation can be blocked by obstructions, such as buildings and trees. The transmission of sunlight is subject to atmospheric absorptions which can be estimated by surface atmospheric pressure and Linke turbidity factors

- **Atmosphere conditions**: including cloudiness, local atmospheric pressure, and Linke turbidity factor. Clouds reflect and scatter solar radiation. The local atmospheric pressure and Linke turbidity factors characterize the reduction of solar radiation by absorption and scattering in the atmosphere. Turbidity index is used to quantify the clarity of the atmosphere. The Linke turbidity factor used in this study is defined as the equivalent number of standard atmospheres (dry and clean) that would have the same attenuation strength produced by the real atmosphere to attenuate extraterrestrial radiation. The factor typically ranges from 1 to 10. As shown in Li and Lam (2002), the factor in Hong Kong is found to be below 5.5 for over half of the cloudless days, which indicates that on these days the clear sky conditions in Hong Kong can be defined as between turbid and clear.
- **Street canyon geometry**: including street orientation and SVF. The SVF is the fraction of sky in the upper hemisphere of a street canyon. A combination of SVF, solar path, and shading mask geometry of the street canyon determines the total incoming solar radiation that reaches the bottom of a street canyon. The orientation of a street canyon affects the timing of exposure to direct sunlight, particularly in a high-density environment with high H/W ratio. In a North–South street canyon, the street surface will be exposed to direct sunlight near mid-day but shaded in other times. However, in an East–West street canyon, it is more likely to be exposed all day from the morning to the afternoon in low latitude cities like Hong Kong (Erell et al., 2014). Therefore, the horizon obstructions of street canyons, including buildings and trees, should be described when calculating the solar irradiance at street level.

The three different components of solar radiation at street level are described as follows:

1. **Direct radiation**

    At the top of the atmosphere, the initial solar total intensity is known as the solar constant, which is approximately from 1366 to 1368 W/m^2, which includes all the different types of solar radiation emitted by the Sun (Johnson, 1954). After entering the atmosphere, the solar radiation interacts with the atmosphere and undergoes complex processes, including atmospheric absorption and scattering, before entering the street canyon, the path length of the solar radiation going through the Earth's atmosphere can be calculated using solar zenith angle (SZA). The air mass the radiation passes through is directly proportional to the SZA. At local zenith direction, when the sun is right above (i.e., SZA = 0°), the air mass equals one which is the minimum.

2. **Diffuse radiation**

    Diffuse radiation is the scattered radiation by the atmosphere when sunlight goes through the atmosphere. Two types of scatterings primarily occur in the atmosphere. The first is Rayleigh scattering (Rayleigh, 1881). Rayleigh scattering is mostly the scattered sunlight by gas molecules which have a much smaller size than the wavelength of sunlight. The Rayleigh scattering is close to isotropic. The second is Mie scattering (Thomas & Stamnes, 2002). Mie scattering occurs when there exist larger size particles compared to light wavelength and can be characterized by a strong forward scattering. Amount of aerosol particles are in the troposphere where most of the Mie scattering by these aerosol particles produced. However, geometric scattering may occur when radiation interacts with even larger particles such as water droplets from the cloud (**B3**). The diffuse radiation contains a large proportion of the UV radiation, which plays a significant effect on studying human exposure to solar radiation.

3. **Reflected radiation**

We assume that the effect of reflections by buildings and ground can be neglected since their effect has a smaller order of magnitude with respect to the beam irradiance in clear sky conditions. In cloudy conditions, such an assumption is not always true since the radiation reflected by urban materials, such as building or ground surfaces, may become a significant part of the total irradiance. As discussed in Sect. 5.4.2, the impact of these reflections maybe comparable to diffused radiation from the atmosphere. This impact needs further investigations using complex 3-D radiative transfer simulations, which, however, are not within the scope of this article. Please refer to Sect. 5.4.2 for further discussions on the reflected radiation in a street canyon and its impact.

2.3.2 Methods of Estimating Solar Radiation

Methods for estimating solar radiation of street canyons can be grouped into three types:

Model Simulation

Model simulations based on 3-D GIS city models, DEM, and solar radiation calculation models for different purposes in urban areas have been reported in previous studies. These studies range from radiation flux simulation (Lindberg & Grimmond, 2011), thermal radiation simulation (Lagouarde et al., 2010), potential photovoltaic usage of solar energy (Hofierka & Kaňuk, 2009), solar radiation absorption by buildings (Terjung & Louie, 1973). Some related publicly available software applications are also commonly used: such as the RayMan model (Matzarakis et al., 2010) and SOLWEIG model (Lindberg et al., 2008). These models allow the parameterization of urban street environment and quantification of street-level incident solar radiation.

However, for quantifying solar radiation in urban environment, most model-based methods require 3-D GIS building model in cities (Fredrik, 2007; Hofierka & Zlocha, 2012; Jakubiec & Reinhart, 2013; Rode et al., 2014) or street photographic descriptions for each location in the model (Fröhlich & Matzarakis, 2013; Matzarakis et al., 2007). Both are not cost-effective and not easy to acquire, especially when there is a need for models of large cities. The street environment in such cities as Hong Kong can be very complex and difficult to be captured by models. In particular, the street tree canopy, a major component of streetscapes, is hard to be parameterized in models.

The Real-Time Measurement from Meteorological Sites

Current meteorological sites for measuring solar radiation provides accurate observation of solar radiation, but they are very sparsely located [e.g., Hong Kong Observatory (2003a, b, c, 2010, 2012, 2016, 2018)]. Moreover, the measurements of global radiation generally refer to an unobstructed horizontal area (e.g., radiation received upon a flat rooftop). Most meteorological studies describing solar radiation refers to measurements or observations taken at free-horizon areas. However, such data fail to give a realistic assessment of the exposures that may be gained at street canyons, where most biological human exposure occurs. A realistic representation of the street-level urban environment, especially in the complex high-density environment, is absent for solar radiation studies.

Street-Sensing Method Based on Street View Panoramas

In a high-density urban environment, solar radiation goes into street canyons through their sky opening, while buildings and trees are the two main obstructions of the solar light path that prevents direct sunlight from reaching the ground at particular times of the day. Since GSV images provide a direct mapping of street morphologies, they can be used to quantify the sky opening and the obstructions by building and trees. Moreover, the high-density urban street morphologies, which are usually complex and have large spatial variations, are not well captured by models but can be fully characterized by GSV images. Therefore, extending from the previous work (Gong, 2019; Gong et al., 2018), this book further proposes using the street geometries characterized based on the street-sensing method, i.e., Google Street View images to quantify the street-level solar radiation in high-density urban areas of Hong Kong (Gong et al., 2019; Yang & Gong, 2025).

2.4 Summary

This chapter describes the theoretical background and physical basis of street-level view factors and solar radiation as well as a literature review the previous methods of quantifying view factors and solar radiation. The main summary is as follows:

1. The previous study areas mainly focus on the cities where streetscape features are relatively simple (compared to high-density urban areas of Hong Kong in this study) with well-defined building and street structures. The feasibility and uncertainty of using Google Street View for estimating SVF in such high-density context are still not clear. Moreover, other street view factors, including 3-D tree canopy and buildings have not been effectively and accurately quantified using street view images. An effective and accurate method for mapping the view factors of the street canyon in Hong Kong is crucial for studying its urban climate and assessing the relevant outdoor thermal comfort.

2. Effective quantification of street-level solar radiation at large-area urban environments is still lacking, because of the following three reasons: (a) Current meteorological sites for measuring solar radiation are very sparsely located and almost all of them are under a full-sky view in free horizon (e.g., Hong Kong Observatory); (b) Model simulations, which are commonly used for describing solar irradiance under urban geometries, rely heavily on expensive 3-D models of cities. In addition, the street environment in such cities as Hong Kong can be very complex and difficult to be captured by models (Gong et al., 2018); (c) User-made photography, such as fisheye images, can be in good details but is only available at a limited number of locations.

Therefore, state-of-the-art tools for quantifying accurately the large-scale urban morphology and solar radiation at the street level is necessary to develop for better urban environment improvement. In the following Chap. 3, the detailed methodology, Google Street View (GSV)-based estimation methods will be presented that describe the modeling process of view factors and solar radiations of street canyons.

References

Abhijith, K. V., & Gokhale, S. (2015). Passive control potentials of trees and on-street parked cars in reduction of air pollution exposure in urban street canyons. *Environmental Pollution, 204*(Supplement C), 99–108. https://doi.org/10.1016/j.envpol.2015.04.013

Akbari, H., Pomerantz, M., & Taha, H. (2001). Cool surfaces and shade trees to reduce energy use and improve air quality in urban areas. *Solar Energy, 70*(3), 295–310. https://doi.org/10.1016/S0038-092X(00)00089-X

Anderson, M. C. (1964). Studies of the woodland light climate: I. The photographic computation of light conditions. *Journal of Ecology, 52*(1), 27–41. https://doi.org/10.2307/2257780

Bourbia, F., & Boucheriba, F. (2010). Impact of street design on urban microclimate for semi arid climate (Constantine). *Renewable Energy, 35*(2), 343–347. https://doi.org/10.1016/j.renene.2009.07.017

Carrasco-Hernandez, R., Smedley, A. R. D., & Webb, A. R. (2015). Using urban canyon geometries obtained from Google street view for atmospheric studies: Potential applications in the calculation of street level total shortwave irradiances. *Energy and Buildings, 86*(Supplement C), 340–348. https://doi.org/10.1016/j.enbuild.2014.10.001

Chen, L., Ng, E., An, X., Ren, C., Lee, M., Wang, U., & He, Z. (2012). Sky view factor analysis of street canyons and its implications for daytime intra-urban air temperature differentials in high-rise, high-density urban areas of Hong Kong: A GIS-based simulation approach. *International Journal of Climatology, 32*(1), 121–136. https://doi.org/10.1002/joc.2243

Erell, E., Pearlmutter, D., Boneh, D., & Kutiel, P. B. (2014). Effect of high-albedo materials on pedestrian heat stress in urban street canyons. *Urban Climate, 10*, 367–386. https://doi.org/10.1016/j.uclim.2013.10.005

Farrar, M. D., Webb, A. R., Kift, R., Durkin, M. T., Allan, D., Herbert, A., & Rhodes, L. E. (2013). Efficacy of a dose range of simulated sunlight exposures in raising vitamin D status in south Asian adults: Implications for targeted guidance on sun exposure. *The American Journal of Clinical Nutrition, 97*(6), 1210–1216. https://doi.org/10.3945/ajcn.112.052639

Fernando, H. J. (2012). *Handbook of environmental fluid dynamics, volume two: Systems, pollution, modeling, and measurements* (Vol. 2). CRC Press.

Fredrik, L. (2007). Modelling the urban climate using a local governmental geo-database. *Meteorological Applications, 14*(3), 263–273. https://doi.org/10.1002/met.29

Fröhlich, D., & Matzarakis, A. (2013). Modeling of changes in human thermal bioclimate resulting from changes in urban design: Example based on a popular place in Freiburg, Southwest Germany. In C. G. Helmis & P. T. Nastos (Eds.), *Advances in meteorology, climatology and atmospheric physics* (pp. 443–449). Springer.

Gál, T., Lindberg, F., & Unger, J. (2009). Computing continuous sky view factors using 3D urban raster and vector databases: Comparison and application to urban climate. *Theoretical and Applied Climatology, 95*(1–2), 111–123. https://doi.org/10.1007/s00704-007-0362-9

Gong, F.-Y. (2019). *Mapping street canyon morphology and solar radiation in high-density urban environments using street sensing approach.* The Chinese University of Hong Kong.

Gong, F.-Y., Zeng, Z.-C., Ng, E., & Norford, L. K. (2019). Spatiotemporal patterns of street-level solar radiation estimated using Google street view in a high-density urban environment. *Building and Environment, 148*, 547–566. https://doi.org/10.1016/j.buildenv.2018.10.025

Gong, F.-Y., Zeng, Z.-C., Zhang, F., Li, X., Ng, E., & Norford, L. K. (2018). Mapping sky, tree, and building view factors of street canyons in a high-density urban environment. *Building and Environment, 134*, 155–167. https://doi.org/10.1016/j.buildenv.2018.02.042

Henderson-Sellers, A. (1995). *Future climates of the world.* Elsevier.

Hofierka, J., & Kaňuk, J. (2009). Assessment of photovoltaic potential in urban areas using open-source solar radiation tools. *Renewable Energy, 34*(10), 2206–2214. https://doi.org/10.1016/j.renene.2009.02.021

Hofierka, J., & Zlocha, M. (2012). A new 3-D solar radiation model for 3-D city models. *Transactions in GIS, 16*(5), 681–690. https://doi.org/10.1111/j.1467-9671.2012.01337.x

Hong Kong Observatory. (2003a). *24-hour time series of mean sea level pressure in Hong Kong.* Retrieved May 13, 2018, from http://www.hko.gov.hk/wxinfo/ts/display_element_pp_e.htm

Hong Kong Observatory. (2003b). *24-hour time series of solar radiation.* Retrieved May 12, 2018, from http://www.hko.gov.hk/wxinfo/ts/display_element_solar_e.htm

Hong Kong Observatory. (2003c). *King's park meteorological station.* Retrieved March 21, 2018, from http://www.hko.gov.hk/wxinfo/aws/kpinfo.htm

Hong Kong Observatory. (2010). *Climate of Hong Kong.* Retrieved June 22, 2018, from http://www.weather.gov.hk/cis/climahk_e.htm

Hong Kong Observatory. (2012). *Direct and diffuse solar radiation information added to Observatory's website.* Retrieved February 22, 2018, from http://www.hko.gov.hk/press/D4/pre20100401e.htm

Hong Kong Observatory. (2016). *The year's weather - 2016.* Retrieved November 27, 2017, from http://www.hko.gov.hk/wxinfo/pastwx/2016/ywx2016.htm

Hong Kong Observatory. (2018). *Climate change in Hong Kong - Cloud amount, solar radiation and evaporation.* Retrieved May 24, 2018, from http://www.hko.gov.hk/climate_change/obs_hk_cloud_e.htm

Jakubiec, J. A., & Reinhart, C. F. (2013). A method for predicting city-wide electricity gains from photovoltaic panels based on LiDAR and GIS data combined with hourly Daysim simulations. *Solar Energy, 93*, 127–143. https://doi.org/10.1016/j.solener.2013.03.022

Johansson, E. (2006). Influence of urban geometry on outdoor thermal comfort in a hot dry climate: A study in fez, Morocco. *Building and Environment, 41*(10), 1326–1338. https://doi.org/10.1016/j.buildenv.2005.05.022

Johnson, F. S. (1954). The solar constant. *Journal of Meteorology, 11*(6), 431–439. https://doi.org/10.1175/1520-0469(1954)011<0431:TSC>2.0.CO;2

Kardakos, E. G., Alexiadis, M. C., Vagropoulos, S. I., Simoglou, C. K., Biskas, P. N., & Bakirtzis, A. G. (2013). Application of time series and artificial neural network models in short-term forecasting of PV power generation. In *Power engineering conference (UPEC), 2013 48th international universities* (pp. 1–6). UPEC. https://doi.org/10.1109/UPEC.2013.6714975

Klemm, W., Heusinkveld, B. G., Lenzholzer, S., & van Hove, B. (2015). Street greenery and its physical and psychological impact on thermal comfort. *Landscape and Urban Planning, 138*(Supplement C), 87–98. https://doi.org/10.1016/j.landurbplan.2015.02.009

Krüger, E. L., Minella, F. O., & Rasia, F. (2011). Impact of urban geometry on outdoor thermal comfort and air quality from field measurements in Curitiba, Brazil. *Building and Environment, 46*(3), 621–634. https://doi.org/10.1016/j.buildenv.2010.09.006

Lagouarde, J.-P., Hénon, A., Kurz, B., Moreau, P., Irvine, M., Voogt, J., & Mestayer, P. (2010). Modelling daytime thermal infrared directional anisotropy over Toulouse city Centre. *Remote Sensing of Environment, 114*(1), 87–105. https://doi.org/10.1016/j.rse.2009.08.012

Li, D. H. W., & Lam, J. C. (2002). A study of atmospheric turbidity for Hong Kong. *Renewable Energy, 25*(1), 1–13. https://doi.org/10.1016/S0960-1481(01)00008-8

Li, X., Ratti, C., & Seiferling, I. (2018). Quantifying the shade provision of street trees in urban landscape: A case study in Boston, USA, using Google street view. *Landscape and Urban Planning, 169*(Supplement C), 81–91. https://doi.org/10.1016/j.landurbplan.2017.08.011

Li, X.-X., & Norford, L. K. (2016). Evaluation of cool roof and vegetations in mitigating urban heat Island in a tropical city, Singapore. *Urban Climate, 16*(Supplement C), 59–74. https://doi.org/10.1016/j.uclim.2015.12.002

Liang, J., Gong, J., Sun, J., Zhou, J., Li, W., Li, Y., & Shen, S. (2017). Automatic sky view factor estimation from street view photographs—A big data approach. *Remote Sensing, 9*(5), 411. https://doi.org/10.3390/rs9050411

Lindberg, F., & Grimmond, C. S. B. (2011). Nature of vegetation and building morphology characteristics across a city: Influence on shadow patterns and mean radiant temperatures in London. *Urban Ecosystems, 14*(4), 617–634. https://doi.org/10.1007/s11252-011-0184-5

Lindberg, F., Holmer, B., & Thorsson, S. (2008). SOLWEIG 1.0—Modelling spatial variations of 3D radiant fluxes and mean radiant temperature in complex urban settings. *International Journal of Biometeorology, 52*(7), 697–713. https://doi.org/10.1007/s00484-008-0162-7

Liou, K. N. (2002). *An introduction to atmospheric radiation*. Elsevier.

Matzarakis, A., Rutz, F., & Mayer, H. (2010). Modelling radiation fluxes in simple and complex environments: Basics of the RayMan model. *International Journal of Biometeorology, 54*(2), 131–139. https://doi.org/10.1007/s00484-009-0261-0

Matzarakis, A., Rutz, F., & Mayer, H. (2007). Modelling radiation fluxes in simple and complex environments—Application of the RayMan model. *International Journal of Biometeorology, 51*(4), 323–334. https://doi.org/10.1007/s00484-006-0061-8

Middel, A., Lukasczyk, J., & Maciejewski, R. (2017). Sky view factors from synthetic fisheye photos for thermal comfort routing—A case study in Phoenix, Arizona. *Urban Planning, 2*(1), 19–30. Retrieved from https://www.cogitatiopress.com/urbanplanning/article/view/855

Morris, C. J. G., Simmonds, I., & Plummer, N. (2001). Quantification of the influences of wind and cloud on the nocturnal urban heat Island of a large city. *Journal of Applied Meteorology, 40*(2), 169–182. https://doi.org/10.1175/1520-0450(2001)040<0169:QOTIOW>2.0.CO;2

Ng, E., Chen, L., Wang, Y., & Yuan, C. (2012). A study on the cooling effects of greening in a high-density city: An experience from Hong Kong. *Building and Environment, 47*(Supplement C), 256–271. https://doi.org/10.1016/j.buildenv.2011.07.014

Oke, T. R. (1981). Canyon geometry and the nocturnal urban heat Island: Comparison of scale model and field observations. *Journal of Climatology, 1*(3), 237–254. https://doi.org/10.1002/joc.3370010304

Oke, T. R. (1982). The energetic basis of the urban heat Island. *Quarterly Journal of the Royal Meteorological Society, 108*(455), 1–24. https://doi.org/10.1002/qj.49710845502

Oke, T. R. (1987). *Boundary layer climates*. Routledge.

Oke, T. R. (1988a). Street design and urban canopy layer climate. *Energy and Buildings, 11*(1), 103–113. https://doi.org/10.1016/0378-7788(88)90026-6

Oke, T. R. (1988b). The urban energy balance. *Progress in Physical Geography: Earth and Environment, 12*(4), 471–508. https://doi.org/10.1177/030913338801200401

Oke, T. R. (1989). The micrometeorology of the urban forest. *Philosophical Transactions of the Royal Society of London, 324*(1223), 335–349. https://doi.org/10.1098/rstb.1989.0051

Oleson, K. (2011). Contrasts between urban and rural climate in CCSM4 CMIP5 climate change scenarios. *Journal of Climate, 25*(5), 1390–1412. https://doi.org/10.1175/JCLI-D-11-00098.1

Oliveira, S., Andrade, H., & Vaz, T. (2011). The cooling effect of green spaces as a contribution to the mitigation of urban heat: A case study in Lisbon. *Building and Environment, 46*(11), 2186–2194. https://doi.org/10.1016/j.buildenv.2011.04.034

Parisi, A. V., Kimlin, M. G., Wong, J. C. F., & Wilson, M. (2000). Diffuse component of solar ultraviolet radiation in tree shade. *Journal of Photochemistry and Photobiology B: Biology, 54*(2–3), 116–120. https://doi.org/10.1016/S1011-1344(00)00003-8

Ratti, C., & Richens, P. (2004). Raster analysis of urban form. *Environment and Planning B: Planning and Design, 31*(2), 297–309. https://doi.org/10.1068/b2665

Rayleigh, L. (1881). On the electromagnetic theory of light. *The London, Edinburgh, and Dublin Philosophical Magazine and Journal of Science, 12*(73), 81–101. https://doi.org/10.1080/14786448108627074

Rhodes, L. E., Webb, A. R., Fraser, H. I., Kift, R., Durkin, M. T., Allan, D., & Berry, J. L. (2010). Recommended summer sunlight exposure levels can produce sufficient ($\geq$20ng ml^{-1}) but not the proposed optimal ($\geq$32ng ml^{-1}) 25(OH)D levels at UK latitudes. *Journal of Investigative Dermatology, 130*(5), 1411–1418. https://doi.org/10.1038/jid.2009.417

Richards, D. R., & Edwards, P. J. (2017). Quantifying street tree regulating ecosystem services using Google street view. *Ecological Indicators, 77*(Supplement C), 31–40. https://doi.org/10.1016/j.ecolind.2017.01.028

Rode, P., Keim, C., Robazza, G., Viejo, P., & Schofield, J. (2014). Cities and energy: Urban morphology and residential heat-energy demand. *Environment and Planning B: Planning and Design, 41*(1), 138–162. https://doi.org/10.1068/b39065

Ross, J. (1981). Incident radiation. In J. Ross (Ed.), *The radiation regime and architecture of plant stands* (pp. 159–174). Springer. https://doi.org/10.1007/978-94-009-8647-3_10

Roth, M., & Chow, W. T. L. (2012). A historical review and assessment of urban heat Island research in Singapore. *Singapore Journal of Tropical Geography, 33*(3), 381–397. https://doi.org/10.1111/sjtg.12003

Sanusi, R., Johnstone, D., May, P., & Livesley, S. J. (2016). Street orientation and side of the street greatly influence the microclimatic benefits street trees can provide in summer. *Journal of Environmental Quality, 45*(1), 167–174. https://doi.org/10.2134/jeq2015.01.0039

Shashua-Bar, L., Tsiros, I. X., & Hoffman, M. E. (2010). A modeling study for evaluating passive cooling scenarios in urban streets with trees. Case study: Athens, Greece. *Building and Environment, 45*(12), 2798–2807. https://doi.org/10.1016/j.buildenv.2010.06.008

Shashua-Bar, L., & Hoffman, M. E. (2002). The green CTTC model for predicting the air temperature in small urban wooded sites. *Building and Environment, 37*(12), 1279–1288. https://doi.org/10.1016/S0360-1323(01)00120-2

Steyn, D. G. (1980). The calculation of view factors from fisheye-lens photographs: Research note. *Atmosphere-Ocean, 18*(3), 254–258. https://doi.org/10.1080/07055900.1980.9649091

Strømann-Andersen, J., & Sattrup, P. A. (2011). The urban canyon and building energy use: Urban density versus daylight and passive solar gains. *Energy and Buildings, 43*(8), 2011–2020. https://doi.org/10.1016/j.enbuild.2011.04.007

Sutherland, R. A., & Bartholic, J. F. (1977). Significance of vegetation in interpreting thermal radiation from a terrestrial surface. *Journal of Applied Meteorology*, (1962–1982), 759–763. https://www.jstor.org/stable/26178158

Tang, U. W., & Wang, Z. S. (2007). Influences of urban forms on traffic-induced noise and air pollution: Results from a modelling system. *Environmental Modelling & Software, 22*(12), 1750–1764. https://doi.org/10.1016/j.envsoft.2007.02.003

Terjung, W. H., & Louie, S. S.-F. (1973). Solar radiation and urban heat islands. *Annals of the Association of American Geographers, 63*(2), 181–207. https://doi.org/10.1111/j.1467-8306.1973.tb00918.x

Thomas, G. E., & Stamnes, K. (2002). *Radiative transfer in the atmosphere and ocean.* Cambridge University Press.

Thorsson, S., Lindberg, F., Björklund, J., Holmer, B., & Rayner, D. (2010). Potential changes in outdoor thermal comfort conditions in Gothenburg, Sweden due to climate change: The influence of urban geometry. *International Journal of Climatology, 31*(2), 324–335. https://doi.org/10.1002/joc.2231

Unger, J. (2004). Intra-urban relationship between surface geometry and urban heat Island: Review and new approach. *Climate Research, 27*(3), 253–264. Retrieved from http://www.jstor.org/stable/24868753

Watson, I. D., & Johnson, G. T. (1987). Graphical estimation of sky view-factors in urban environments. *Journal of Climatology, 7*(2), 193–197. https://doi.org/10.1002/joc.3370070210

Webb, A. R. (2006). Considerations for lighting in the built environment: Non-visual effects of light. *Energy and Buildings, 38*(7), 721–727. https://doi.org/10.1016/j.enbuild.2006.03.004

Wong, N. H., & Yu, C. (2005). Study of green areas and urban heat Island in a tropical city. *Habitat International, 29*(3), 547–558. https://doi.org/10.1016/j.habitatint.2004.04.008

Yang, Z., & Gong, F.-Y. (2025). Utilizing street view images to estimate solar energy potential for photovoltaic-powered buses. *Applied Geography, 177*, 103567. https://doi.org/10.1016/j.apgeog.2025.103567

Chapter 3
Methodological Innovations in Urban Canyon Analysis

Contents

3.1	Overview.	30
3.2	Study Area and Data Collection.	30
	3.2.1 Study Area.	30
	3.2.2 Data Collection.	32
3.3	GSV-Based Estimation of Street-Level View Factors.	35
	3.3.1 Collecting GSV Panorama Images.	35
	3.3.2 Extractions of Street Features Using Deep-Learning Techniques.	37
	3.3.3 Projection Into Fisheye Images and Calculations of View Factors.	38
3.4	GSV-Based Estimation of Street-Level Solar Radiation.	39
	3.4.1 Attributes Collection and Features Extractions from GSV Images.	40
	3.4.2 Urban Canyon Geometry Calculation Using GSV Images and Solar Path.	42
	3.4.3 Calculation of Street-Level Solar Radiation.	43
3.5	Summary.	48
References.		49

Abstract This chapter outlines the theoretical and methodological frameworks for analyzing urban street canyon morphology and its role in regulating solar radiation at street level. It introduces critical geometric indicators—sky view factor (SVF), tree view factor (TVF), and building view factor (BVF)—to quantify urban radiative environments, emphasizing their relevance to microclimate studies, UHI mitigation, and thermal comfort. Traditional quantification methods, including 3D modeling and fisheye photography, are contrasted with emerging street-sensing techniques using google street view (GSV) panoramas, which offer scalable, high-resolution urban mapping. This chapter delineates the physical components of street-level solar radiation (direct, diffuse, and reflected) and their dependence on solar geometry, atmospheric turbidity, and canyon structure. Challenges in high-density cities, such as Hong Kong, are highlighted, where complex geometries and vegetation obstruct conventional modeling. A novel GSV-based approach is proposed to overcome limitations in existing methods, enabling precise, large-scale quantification of view factors, and solar irradiance. By integrating computational tools like pyramid scene parsing networks (PSPNet), this methodology enhances urban climate assessments, supporting sustainable planning and heat resilience strategies. The discussion underscores the necessity of accurate street-level radiation modeling for improving human health outcomes and urban ecosystem services in rapidly urbanizing environments.

Keywords Google street view (GSV) · View factors · Solar radiation estimation · High-density urban areas · Deep learning

3.1 Overview

In this chapter, the methodology of google street view (GSV)-based street view factors and solar radiation estimations will have illustrated. The study will use the high-density urban areas of Hong Kong, i.e., Kowloon area and Hong Kong Island, as the study area. This chapter illustrates the sky, tree, and building view factors and the spatiotemporal patterns of street-level solar energy estimated using Google Street View images high-density urban area of Hong Kong. The methodological framework mainly made up two board parts:

- Part I develop an approach for estimating and mapping view factors inducing openness (sky view factor), tree canopy (tree view factor), and building density (building view factor) of street canyons using Google Street View images in a complex urban living environment context (i.e., Hong Kong). To further conduct the spatial patterns of sky, tree, and building view factor of the study area.
- Part II further develop an approach for estimating and mapping solar radiation of street canyons with complex urban living environment context using Google Street View images. To further conduct the spatial pattern and temporal variation of street-level solar irradiance of the study area.

3.2 Study Area and Data Collection

3.2.1 Study Area

Hong Kong, situated at the coastline of southeastern China (see Fig. 3.1a), is one of the most densely-populated and built-up cities in the world. It has a population of over seven million living in around 262 km^2 of developed land (Census and Statistics Department, The Government of Hong Kong S A R, 2016; Planning Department, The Government of the Hong Kong, S A R, 2016). The climate of

3.2 Study Area and Data Collection

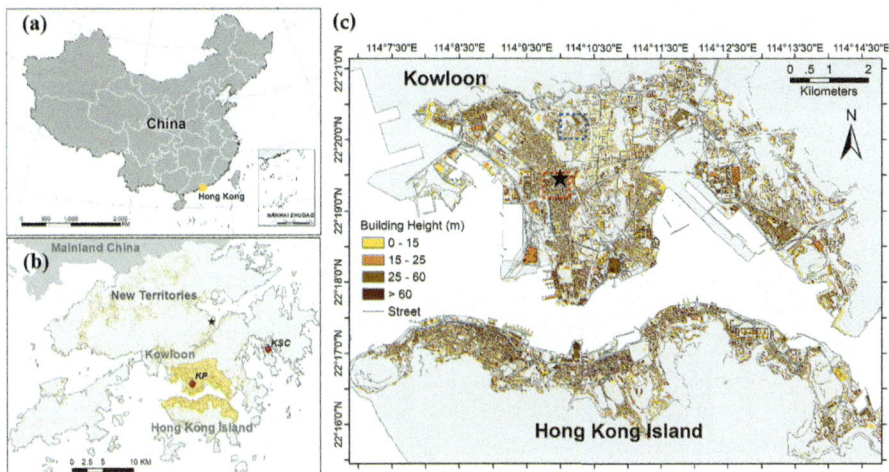

Fig. 3.1 (**a**) Location of Hong Kong (yellow circle) in southeastern China; (**b**) High-density urban areas in Hong Kong, as outlined in yellow, including Kowloon and northern Hong Kong Island; the yellow points are the spatial destruction of building footprints; the red points are the locations of King's Park and Kau Sai Chow sites from Hong Kong Observatory; the black point is the location of field measurement of in our study (one street canyon of the campus at the urban area of New Territories); (**c**) building density map, including distribution and height, overlaid with streets in gray. The building and street data are extracted from the B5000 maps series by the Hong Kong Lands Department. The dotted red and blue rectangles outline the field survey regions for high-rise and low-rise regions, respectively, as described in Sect. 4.2.2. The black star is the location of the street canyon example in Mong Kok shown in Fig. 3.2

Hong Kong is subtropical maritime, which features hot and humid summers and warm winters (Hong Kong Observatory, 2016). Moreover, high-density urban areas of Hong Kong are characterized by high-rise compact building blocks and deep street canyons with a high H/W ratio. In these areas, tall buildings of some 40–60 stories lining narrow streets of 15–25 m width have been the norm. Serious issues related to human thermal comfort (Ng & Cheng, 2012), air pollution (Lu et al., 2011), and the UHI effect (Wang et al., 2016) due to its climate and urban morphologies have been primary planning concerns. As effective indicators for characterizing urban streetscapes, street VFs have been widely incorporated in modeling to address these concerns. However, accurate assessment of VF estimations, which is crucial for quantifying the uncertainty of models, is still lacking due to a lack of measurements.

In this study, high-density urban regions of Kowloon and Hong Kong Island are chosen as our study area, as shown in Fig. 3.1b. This area is one of the most densely built and populated areas in the world, with an average building height of 27 m with a standard deviation of 30.7 m and a population density of around 42,900 persons per km^2. As shown in Fig. 3.1c, the building heights are grouped into high rise (>25 m), mid-rise (15–25 m), and low rise (<15 m), according to local climate zone classification in Hong Kong urban areas (Stewart & Oke, 2012; Zheng et al., 2017).

Fig. 3.2 An example of the deep street canyon in the Mong Kok area (the black star shown in Fig. 3.1), one of the typical high-density high-rise urban areas of Hong Kong (Google Street View, 2017)

Most high-rise buildings are distributed in southern Kowloon and northern Hong Kong Island.

A typical street in high-density urban areas of Hong Kong is characterized by high-rise buildings, narrow compacted streets, interferences of heavy travel volume and pedestrian flow, and complex streetscapes with the amount of colorful overhanging signboards that block sunlight and air paths and provide limited openness to the sky (see Fig. 3.2). An effective and accurate method for mapping the VFs of the street canyon in Hong Kong is therefore crucial for studying its urban climate and assessing the relevant outdoor thermal comfort. General characters of urban geometry in Hong Kong are as follows:

(a) High-density high-rise constructions;
(b) Vertical surface of building materials with high heat storage;
(c) Ground surface paved with high-impervious materials, i.e., asphalt, cement, and concrete;
(d) The block urban geometry trapping the radiation and air ventilation;
(e) The limited vegetation density at microscale environment.
(f) High energy and waste release from buildings.

3.2.2 Data Collection

Google Street View serves millions of Google users daily with panorama images captured in hundreds of cities in 20 countries across four continents (Anguelov et al., 2010), which are freely accessible on Google Maps by the Google Street View API (Google Maps APIs, 2017a, 2017b). Figure 3.3 shows the Google Street View

3.2 Study Area and Data Collection

Fig. 3.3 Google Street View coverage 3-D map of Kowloon Area (Google Street View, 2017)

coverage 3-D map of Kowloon peninsula, in which the streets with street view images available are shown as blue lines on Google Maps.

In this study, we use publicly accessible GSV images to estimate the SVF, TVF, and BVF of street canyons in high-density urban areas of Hong Kong. Street panorama images sampled at 30-m intervals are first collected using the GSV API (Google Maps APIs, 2017a, 2017b) based on the latitudes and longitudes of the sampling points. Extraction of features, including sky, trees, and buildings, is implemented using the scene parsing method in a deep-learning framework (Zhao et al., 2016; Zhou et al., 2016).

For GSV-based view factors estimation in Part I, street panorama images sampled at 30-m intervals are first collected using the GSV Application Programming Interface (API) (Google Maps APIs, 2017a, 2017b) according to the location (latitude and longitude), horizontal field of view, compass heading, and the vertical angle of the camera relative to the street view vehicle (Google Maps APIs, 2017c). A total of 26 (width) times 13 (height) tiles are obtained and combined to get a complete panorama image. A total of 33,544 images are collected in the study area. Invalid GSV images, including those with empty content, are filtered out. Example of GSV images of a typical high-density street canyon is shown in Fig. 3.2.

For GSV-based solar radiation estimation in Part II, the street orientation is not relevant to view factor calculations, the heading of the panorama image is important information to be extracted from the GSV for determining the street geometries and overlaid solar path. By default, the panorama image obtained from combining tiles centers in the vehicle heading direction, which can be requested from the Street

View Service providing the latitude and longitude of the panorama image (Google Maps APIs, 2017c). Eventually, the center heading of the panorama images can be adjusted to the north by shifting the corresponding vehicle heading direction in degrees. The calculation of street view factors in this study refers to the position of the GSV vehicle moving along the roadway which is slightly different from the road axis. Given that most streets in Hong Kong have either one lane or two lanes (almost all in high-density urban areas), the axis of the road and the vehicle path are therefore close to each other. So, we assume that the calculation of solar irradiance using GSV is representative for the street axis.

To calculate the solar irradiance, the following meteorological data are obtained:

(a) Solar geometries and extraterrestrial radiation. These two datasets can be calculated by the Solar Position and Intensity (SOLPOS) algorithm developed by the National Renewable Energy Laboratory (2018). The algorithm generates the solar position, including solar zenith angle and solar azimuth angle, and extraterrestrial radiation with small uncertainty based on inputs of location, date, and hour. For each day, solar geometries and extraterrestrial radiation at 10-min interval are obtained;
(b) Sea-level pressure in Hong Kong obtained from mean sea-level pressure observations by Hong Kong Observatory (2003a);
(c) Cloudiness obtained from measurements from HKO; and.
(d) Monthly Linke turbidity factor in Hong Kong are obtained by Li and Lam (2002). Monthly averages of the Linke turbidity factors are used in this study.

Table 3.1 provides the details of these input datasets. The monthly variation of Linke turbidity factors is shown in Fig. 3.4, which shows the monthly averaged Linke turbidity factors (in black) in Hong Kong. These factors are adopted from Table 3.1 in Li and Lam (2002). We calculated the average Linke turbidity factor from three estimates: T_{Lin}, T_{Lou}, and T_{Pin} which correspond to three different estimation methods. The T_{Vis} in Li and Lam (2002) is not used because it is very different from the other three estimates.

Table 3.1 Summary of input parameters, data source, the temporal resolution used in this study, and corresponding dataset descriptions

	Data source	Temporal resolution	Description
(a) Extraterrestrial radiation	National renewable energy laboratory	10 min	Calculated using SOLPOS (National Renewable Energy Laboratory, 2018) given information of location and time
(b) Sea-level pressure	HKO	1 min	Measured at King's Park Meteorological Station by HKO (Hong Kong Observatory, 2003a, 2003c)
(c) Cloudiness	HKO	Hourly	Reported in octas hourly by trained observers at the HKO Headquarters (Hong Kong Observatory, 2018)
(d) Linke turbidity factor	Li and Lam (2002)	Monthly	Averaged from three different estimates

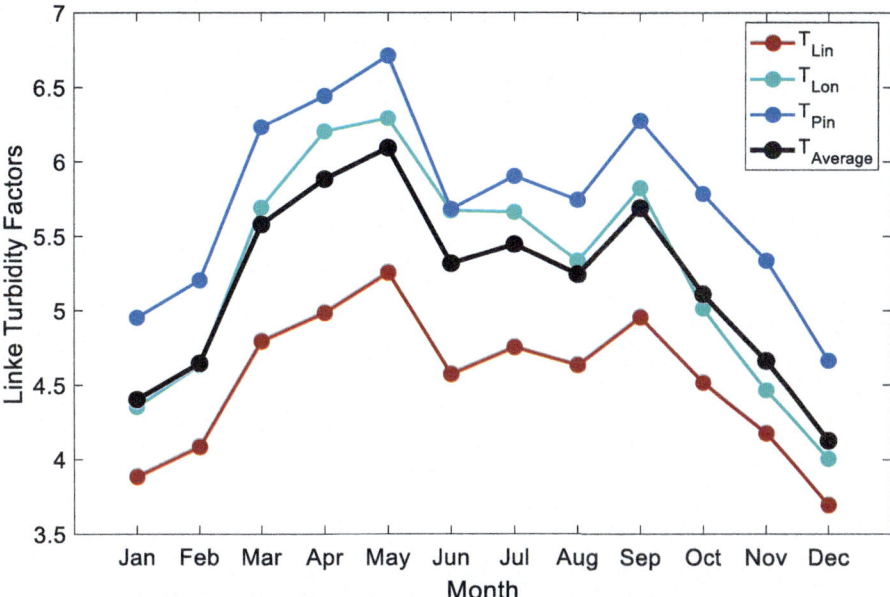

Fig. 3.4 Monthly averaged Linke turbidity factors (in black) in Hong Kong. Based on Table 2 in Li and Lam (2002), the Linke turbidity factor used in this study is the average value of three different estimates: T_{Lin} (in red), T_{Lou} (in green), and T_{Pin} (in blue)

3.3 GSV-Based Estimation of Street-Level View Factors

In this study, we project the panorama images from cylindrical to azimuthal projection to generate the fisheye images. From the fisheye images, VFs are calculated by applying the classical photographic method (Johnson & Watson, 1984). Figure 3.5 shows the workflow procedure for VF calculations using this method, and a detailed description is as follows.

3.3.1 Collecting GSV Panorama Images

We sampled points at 30-m intervals along the street lines, shown in Fig. 3.1c, using GIS software. There is a total of 33,544 sample points in the study area. We collected the GSV panorama images for all the sampled points in the following ways:

1. Obtain panorama image ID at a specific location using the following URL:
    https://maps.googleapis.com/maps/api/streetview/metadata?size=400x400&location=LAT,LON&heading=HEADING&fov=FOV&pitch=PITCH&key=APIKey.

where LAT, LON are the latitude and longitude, respectively, FOV (90 by default) determines the horizontal field of view of the image, HEADING (0 by default) indicates the compass heading of the camera, PITCH (0 by default) specifies the up or down angle of the camera relative to the street view vehicle, and API key is the credential required to authenticate the request.

2. Download tiles of a panorama image using the following URL:
    http://cbk0.google.com/cbk?output=tile&panoid=PANO_ID&zoom=5&x=I&y=J

    where PANO_ID is obtained from the above step; and I (from 0 to 25) and J (from 0 to 12) are the row and column indices of the image tile. We can get a complete panorama image by combining 26 × 13 tiles. Examples of GSV images in high-rise and low-rise areas are shown in Fig. 3.5a.

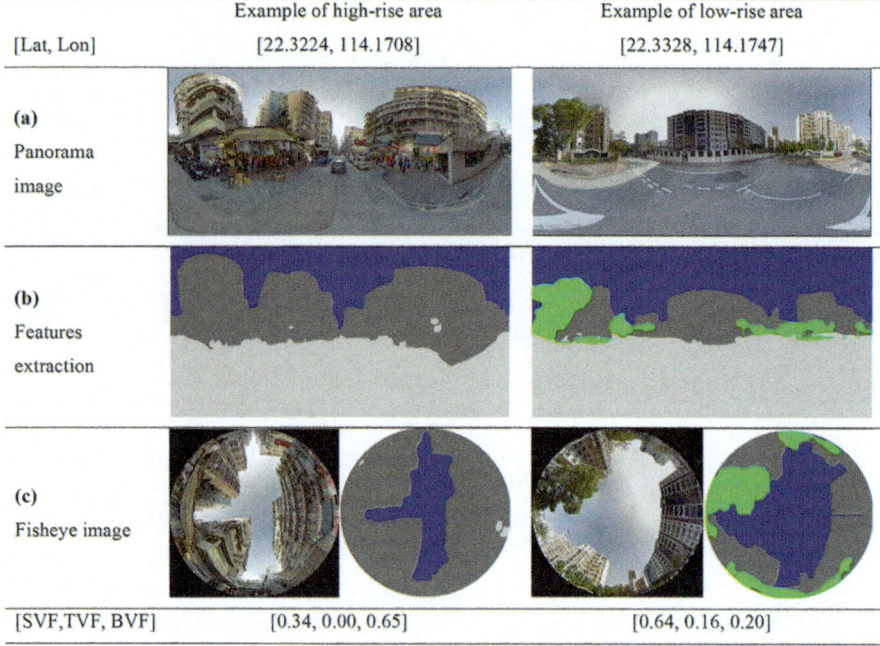

Fig. 3.5 Workflow procedure for VF calculations using GSV images, illustrated by taking two examples from high-rise and low-rise areas. (**a**) Panorama images downloaded from Google servers using coordinates of sampling street points as inputs. (**b**) Extraction of sky (in blue), trees (in green), and buildings (in gray) using the scene parsing deep-learning technique (Zhao et al., 2016). (**c**) Fisheye images obtained by projecting the panorama images from cylindrical projection to azimuthal projection. Based on the fisheye image of extracted features, SVF, TVF, and BVF are calculated using the classical photographic method developed by Johnson and Watson (1984). The resulted VF estimates are also indicated

3.3.2 Extractions of Street Features Using Deep-Learning Techniques

We propose the use of the scene parsing method in a deep-learning framework to extract street features, including sky, trees, and buildings, from the GSV images. The deep-learning model employed in this study is the Pyramid Scene Parsing Network (PSPNet) (Zhao et al., 2016). In essence, scene parsing segments and parses an image into different image regions associated with semantic categories, including sky, trees, and buildings.

Deep learning based on deep convolutional neural network (CNN) allows computational models that are composed of multiple processing layers to learn representations of natural data with multiple levels of abstraction and has been widely used in image classification and pattern recognition (LeCun et al., 2015). Enabled by the proliferation of deep-learning techniques, a number of CNN based model for semantic scene parsing, such as understanding street features, have been proposed and have achieved outstanding performance (Badrinarayanan et al., 2017; Lin et al., 2016; Long et al., 2015). Compared with the previous works, the architecture of PSPNet paid more attention to feature ensembling and structure prediction, to integrate global context information into the prediction process. The PSPNet model employed in this study provides a fully convolutional network (FCN) based pixel prediction framework which is superior for processing difficult scenery context features. It is a practical system for state-of-the-art scene parsing and semantic segmentation where all crucial implementation details are included (Zhao et al., 2016). PSPNet uses a pre-trained semantic segmentation network based on the ADE20K dataset.

ADE20K contains more than 20 K scene-centric images exhaustively annotated with objects and objects parts. Specifically, the benchmark is divided into 20 K images for training, 2 K images for validation, and another batch of held-out images for testing. There are totally 150 semantic categories included for evaluation, which include stuff like sky, road, grass, and move objects like person, car and bus. When evaluating prediction accuracies on various datasets, it achieves state-of-the-art performance and outperforms many other models in the semantic scene parsing frameworks. In particular, it achieves a high accuracy of 80.2% in predicting 150 object classes of cityscapes, a dataset for semantic urban scene understanding collected from 50 cities in different seasons (Zhou et al., 2016).

Figure 3.6 shows the workflow of semantic scene parsing using PSPNet. The downloaded and combined GSV panorama image is first resampled into 473 × 946 pixels. The panorama image is then separated into two images with 473 × 473 pixels, a size as required by the deep-learning module, before consecutively inputting the two images into the PSPNet, as shown in Fig. 3.6a. PSPNet uses CNN to get the feature map of the last convolutional layer as shown in Fig. 3.6b. A pyramid parsing

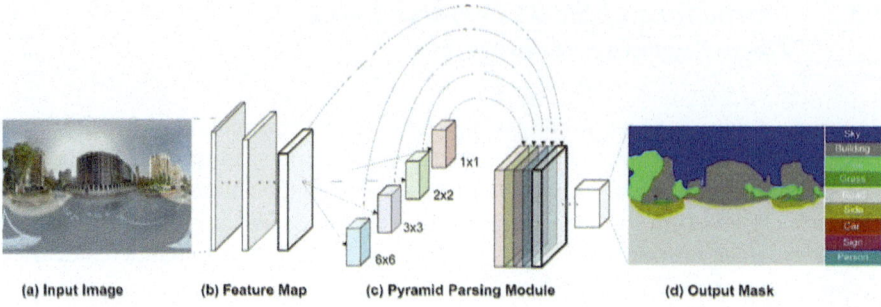

Fig. 3.6 Workflow of semantic scene parsing using PSPNet. For a given input street view image in (**a**), the network extracts the feature map in (**b**), and then the pyramid parsing module is applied to form the final feature representation of the streetscape in (**c**). Finally, a pixel-wise classified output street view image with semantic categories in (**d**) produced by feeding the feature representation into a convolution layer

module is applied to harvest different sub-region representations of the image, followed by upsampling and concatenation layers to form the final feature representation of the streetscape, which carries both local and global context information, as illustrated in Fig. 3.6c. It fuses features under four different pyramid scales with bin sizes of 1×1, 2×2, 3×3, and 6×6, respectively. Finally, the representation is fed into a convolution layer to get the final per-pixel prediction and produce a pixel-wise classified out street view image with semantic categories in Fig. 3.6d. The size of output images is also 473×473 pixels, includes 150 classifications, and features of the sky, trees, and buildings are extracted and calculated in this study. Using a server of a workstation with 8 cores CPU and NVIDIA 1080Ti GPU (12G RAM). It took about 20 h to finish the 33,544 images, roughly about 2 s per panorama image. TVF is general definition vegetation, including grass and trees. Based on the calculation results of 33,544 GSV images using deep learning, both values of SVF and TVF are zero will be filtered, as these points are distributed indoors or in tunnels after checking the original GSV images. The study uses the filtered 29,264 GSV images for further analysis. Examples of the extracted features are shown in Fig. 3.5b.

3.3.3 Projection Into Fisheye Images and Calculations of View Factors

Projection Into Fisheye Images

We use the photographic method which applies a fisheye lens to the panorama image in order to project the hemispheric environment (cylindrical projection) onto a circular plane (azimuthal projection) and generate the fisheye image in the following way (Li et al., 2018). This projection is implemented by constructing a relationship between pixels (x_f, y_f) on a fisheye image and (x_p, y_p) on a panorama image, given by,

$$x_p = \begin{cases} \left(\pi/2 + \tan^{-1}\left[(y_f - C_y)/(x_f - C_x)\right]\right) \times W_p/2\pi, & x_f < C_x \\ \left(3\pi/2 + \tan^{-1}\left[(y_f - C_y)/(x_f - C_x)\right]\right) \times W_p/2\pi, & x_f > C_x \end{cases} \quad (3.1)$$

$$y_p = \left(\sqrt{(x_f - C_x)^2 + (y_f - C_y)^2}/r_0\right) \times H_p \quad (3.2)$$

where W_p and H_p are the width and height of the panorama image, respectively; $r_0 = W_p/2\pi$ is the radius of the fisheye image; and (C_x, C_y) are the coordinates of the center pixel on the fisheye image; $C_x = C_y = W_p/2\pi$. First, an empty fisheye image is initialized with coordinates (x_f, y_f). By using Eqs. (3.1) and (3.2), each pixel with coordinate of (x_f, y_f) is then uniquely connected to a pixel with coordinate of (x_p, y_p) in the panorama image. Lastly, the pixel value in (x_f, y_f) can be assigned by that in (x_p, y_p) and this process is repeated for each pixel until the fisheye image is fully constructed.

Calculations of VFs

To calculate VFs, based on the equation in Johnson and Watson (1984), we divide the fisheye image into a number of concentric annuli of equal width, and then sum up all the annular sections representing the sky, trees, and buildings to calculate the SVF, TVF and BVF, respectively, using the following formula:

$$\Psi_x = \frac{1}{2\pi} \sin\frac{\pi}{2n} \sum_{i=1}^{n} \sin\left[\frac{\pi(2i-1)}{2n}\right] \alpha_{i,x} \quad (3.3)$$

where x can be sky, tree, or building, n is the total number of rings (here we use 100); i (from 1 to 100) is the index of the ring; $a_{i,x}$ is the angular width of pixels of feature x (x can be sky, tree, or building) in the ith ring. Examples of the generated fisheye images are shown in Fig. 3.5c. The SVF, TVF, and BVF quantify the fractions of the sky, trees, and buildings, respectively, of the built environment seen from a particular observation point on the ground within the built environment.

3.4 GSV-Based Estimation of Street-Level Solar Radiation

Part II investigates the street-level solar energy estimated using Google Street View images. The methodology framework of this part is presented in Fig. 3.7 and consists of three main phases. In Phase 1 (green rectangles), the sky, tree and building

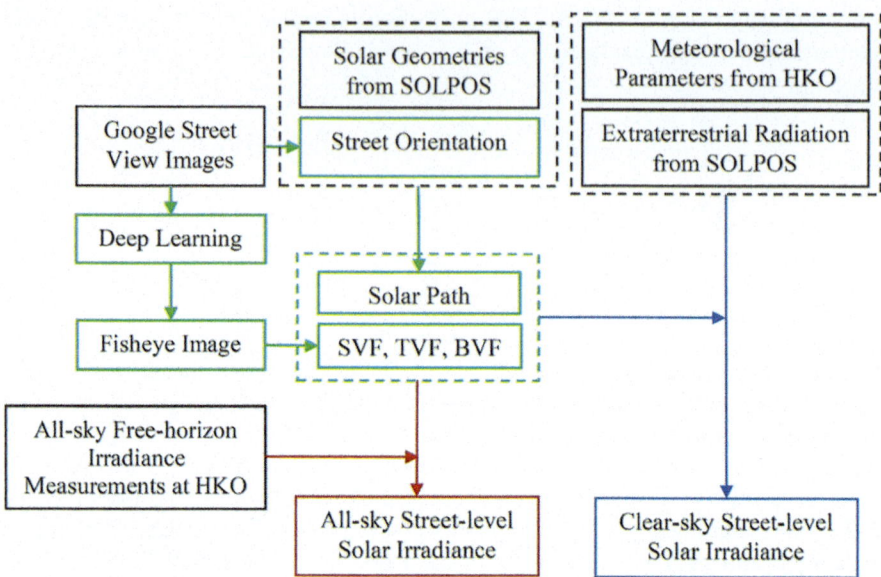

Fig. 3.7 Schematic framework for this study, in which black rectangles represent the collected datasets, green rectangles represent the calculations of solar path and view factors in street canyons using GSV images, the blue rectangle represents the calculation of clear-sky street-level solar irradiance, and the red rectangle represents the calculation of all-sky street-level solar irradiance in the high-density urban areas of Hong Kong

view factors are first estimated from the GSV images based on Part I study (Gong et al., 2018) and the diurnal solar path is calculated based on the street and solar geometries; In Phase 2 (blue rectangle), solar irradiance in clear-sky, free-horizon conditions is first calculated and then combined with street morphologies and the solar path to derive the clear-sky street-level solar irradiance; In Phase 3 (red rectangle), direct and diffuse components measurements from HKO are used as all-sky free-horizon solar irradiance and combined with street morphologies and solar path to derive the all-sky street-level solar irradiance.

3.4.1 Attributes Collection and Features Extractions from GSV Images

In this section, we use publicly accessible GSV images to estimate the solar geometry and radiation of street canyons in high-density urban areas of Hong Kong (Gong et al., 2019; Yang & Gong, 2025). The workflow procedure is shown in Fig. 3.7 with an example of a typical high-density street canyon. Street panorama images sampled at 30-m intervals are first collected using the GSV API (Google Maps APIs, 2017a, 2017b) based on street view locations (the latitudes and

3.4 GSV-Based Estimation of Street-Level Solar Radiation

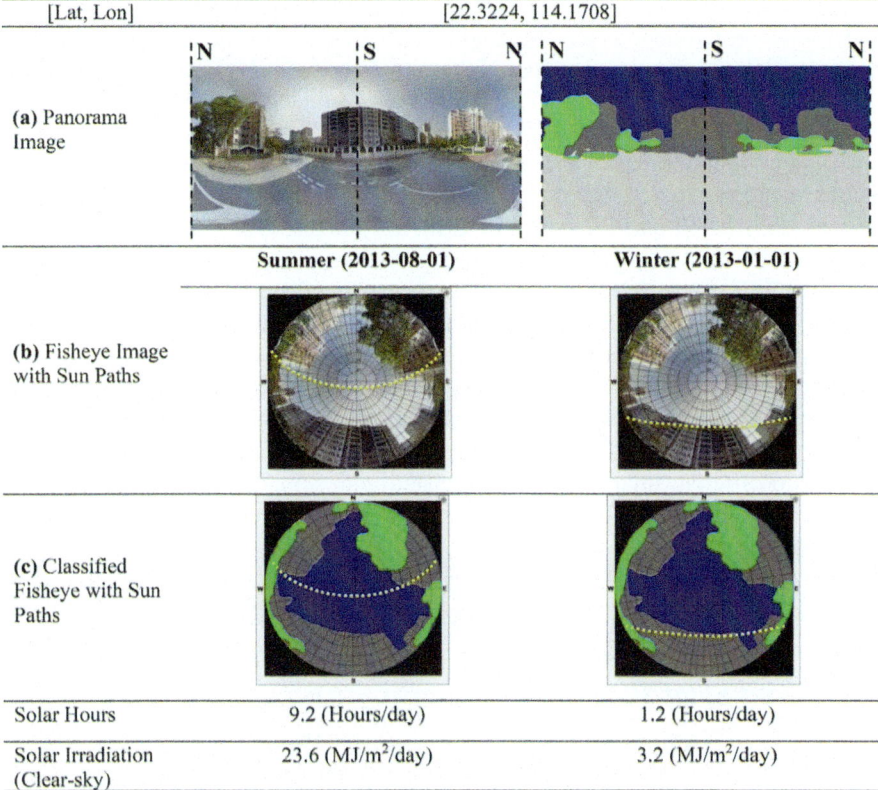

Fig. 3.8 Workflow procedure for solar radiation calculation using GSV image shown using an example of a street canyon in high-density urban areas of Hong Kong. (**a**) Panorama image collected using the GSV API, and extractions of sky (in blue), trees (in green), and buildings (in deep gray) using a deep-learning technique. (**b**) Street geometries, including zenith angle indicated by concentric circles and azimuth angle indicated by radius lines, in a fisheye and the overlaid sun paths of summer (August 1st) and winter (January 1st); (**c**) Same as (**b**) but with classified fisheye images. In addition, solar hour and clear-sky solar radiation are also indicated

longitudes), and point-of-view (POV) (headings) of the sampling points (Google Maps APIs, 2017c). Extraction of solar surfaces, including sky, vegetation (tree, grass, flora, and palm), and building (building, edifice, house, bridge and span), is implemented using the scene parsing method in a deep-learning framework (Zhao et al., 2016; Zhou et al., 2016). We then project the panorama images from cylindrical to azimuthal projection to generate the fisheye images. Figure 3.8 shows the workflow procedure for solar radiation calculation using Google Street View image method, and a detailed description is as follows.

We sampled points at 30-m intervals along the street lines shown in Fig. 3.1c, using GIS software. There is a total of 33,544 sample points in the study area. We collected the GSV panorama images and obtain panorama image ID for all the sampled points. There are five attributes of one sampled point with the specific location, including latitude (LAT), longitude (LON), the horizontal field of view of the image (FOV), the compass heading of the camera (HEADING), and the up or down angle of the camera relative to the street view vehicle (PITCH). One panorama image by combining I × J tiles, where I (from 0 to 25) and J (from 0 to 12) are the rows and column indices of the image tile. Therefore, we can get a complete panorama image by 26 × 13 tiles. Invalid GSV images, including those with empty content, are filtered out. Examples of GSV images one typical high-density street canyon is shown in Fig. 3.8a.

Extractions of the free sky and solid surface using deep-learning techniques. We propose the use of the scene parsing method in a deep-learning framework to extract different solid surfaces, including sky, vegetation (tree, grass, flora, and palm), and building (building, edifice, house, bridge and span), from the GSV images. The deep-learning model employed in this study in the Pyramid Scene Parsing Network (PSPNet) (Zhao et al., 2016). In essence, scene parsing segments and parses an image into different image regions associated with semantic categories, including sky, trees, and buildings. The classification result of features extraction is shown in Fig. 3.8a.

3.4.2 Urban Canyon Geometry Calculation Using GSV Images and Solar Path

To calculate the solar radiation, the following meteorological data are obtained: (a) Solar geometries and extraterrestrial solar radiation in Hong Kong. These two datasets can be calculated by the SOLPOS algorithm developed by the National Renewable Energy Laboratory (2018). The algorithm generates the solar position, including solar zenith angle and solar azimuth angle, and extraterrestrial global horizontal solar energy with small uncertainty based on inputs of location, date, and hour. For each day, solar geometries and extraterrestrial solar radiation at 10-min interval are obtained; (b) Sea-level pressure in Hong Kong obtained from mean sea-level pressure observations by Hong Kong Observatory (2003a); (c) Cloudiness obtained from cloudiness measurements from HKO; (d) Monthly Linke turbidity factor in Hong Kong are obtained from previous study by Li and Lam (2002). Monthly averages of the Linke turbidity factors are used in this study.

The position of the sun at a fixed observation point is normally determined by specifying two angles. Solar zenith angle (φ, 0°–90°) is defined as the angle between the vertical above the observer and the connecting line between the observer and the sun; solar azimuth angle (ψ, 0°–360°) is defined as the angle between the horizon with the value increasing in clockwise direction (North is 0°). According to the solar position algorithm, two sun angles in the polar coordinate calculation can be calculated, based on the date, time, longitude, latitude of each GSV images. The results of the sun path calculation are showed as the yellow dotted lines in Fig. 3.8b, which is the calculation results of sun position and tracking the path of the sun for one

3.4 GSV-Based Estimation of Street-Level Solar Radiation

geographical location of street canyons during each day. The polar coordinates r (the radial coordinate) and θ (the angular coordinate, often called the polar angle) are defined in terms of Cartesian coordinates by

$$x = r\cos\theta$$
$$y = r\sin\theta \qquad (3.4)$$

where r is the radial distance from the origin and θ is the counter clockwise angle from the x-axis.

$$r = \sqrt{x^2 + y^2}$$
$$\theta = \tan^{-1}(y/x) \qquad (3.5)$$

In the polar coordinates of radiation geometry, the value of r standards for Solar Zenith Angle φ (0–90°), and the value of θ standards for Solar Azimuth Angle ψ (0–360°). Each sun position at 10-min interval is located in relevant polar coordinate and conducts the sun path with 144 points totally. Figure 3.8b shows the sun paths of two different seasonal days, i.e., one summer day (August 1, 2013) and one winter day (January 1, 2013). Furthermore, Fig. 3.8c illustrates the calculation results of solar radiation for each street using Google Street View images. Calculate the solar geometries and solar irradiation based on the lat/lon and time. We can get the following output from National Renewable Energy Laboratory (2018): (1) Solar zenith angle (degrees from zenith, refracted); (2) Solar azimuth angle; (3) Solar elevation (no atmospheric correction); (3) Solar elevation angle (degrees from horizon, refracted); (4) Extraterrestrial Global Horizontal Solar Irradiance (W/m²); (5) Extraterrestrial Direct Normal Solar Irradiance (W/m²); (6) Extraterrestrial Global Irradiance on a tilted surface (W/m²); (7) Julian Day of 1 Jan 2000 minus 2,400,000 days. Based on the surface feature classification using deep learning in Sect. 3.3.2, we can get the accurate location of 144 sun position (relative sun time) in different solar surfaces or free sky. Therefore, we can calculate the solar irradiance for one location of the street canyon (see Fig. 3.8c).

3.4.3 Calculation of Street-Level Solar Radiation

As indicated in Fig. 2.1, three key factors influence the street-level solar irradiance in a high-density environment: (1) Solar geometries including the solar zenith angle (SZA) and solar azimuth angle (SAA). In Hong Kong (22°17′7.87″N, 114°9′27.68″E), the SZA at local noon time changes from 45.5° in the winter and 3.5° in the summer. The solar geometries determine the associated solar irradiance on a horizon surface due to seasonal variation; (2) Atmosphere conditions including cloudiness, local atmospheric pressure, and Linke turbidity factor. As shown in Li and Lam (2002), the Linke turbidity factor in Hong Kong is found to be below 5.5 for over half of the cloudless days, which indicates that on these days the clear-sky conditions in Hong Kong can be defined as between turbid and clear; (3) Street canyon geometry including street orientation and SVF. In an East-West street canyon, it is more likely to be exposed all day from the morning to the afternoon in low latitude cities like Hong Kong (Erell et al., 2014). Therefore, the horizon obstructions of street canyons, including buildings and trees, will be described when calculating the solar irradiance at street level.

For practicality, the direct and diffuse components of solar radiation reaching an urban street canyon can be separately approximated to be fractions of those measured in open space under an unobstructed view of the sky. As shown in the comparison of our calculations with field measurements in Sect. 5.2.2, the contribution from the reflection is actually very small compared to the daily solar irradiation.

As shown in the RayMan model (Matzarakis et al., 2007), the incoming solar radiation at the upper hemisphere of a street canyon includes the direct and the diffuse components, as shown in Eq. (3.6). The effect of street canyon geometries on these two components can be considered independently. For cloudless skies, direct irradiance in the street canyon depends on whether the solar radiation is blocked by buildings or trees, while diffuse irradiance is approximately proportional to the SVF.

$$G_{street} = I_{street} + D_{street} \tag{3.6}$$

The direct irradiance is given by,

$$I_{street} = I_{open} \times f \tag{3.7}$$

where I_{open} is the free-horizon direct irradiance and f is a binary indicator of the existence of solar ray path obstructions; f equals zero if the ray path is masked by an obstacle (such as buildings or trees), otherwise f equals one. These obstacles can be extracted using GSV images.

The diffuse irradiance is given by,

$$D_{street} = D_{iso\_open} \times \Psi_{sky} + D_{aniso\_open} \times f + D_{cloud\_open} \times \Psi_{sky} \tag{3.8}$$

where D_{iso_open} is the isotropic diffuse radiation on a horizontal surface in a free horizon; D_{aniso_open} is the anisotropic diffuse radiation on a horizontal surface in a free horizon, which tends to concentrate in the vicinity of the sun and is discriminated depending on whether the sun is directly visible; D_{cloud_open} is the diffuse radiation on a horizontal surface in a free horizon due to clouds; Ψ_{sky} is the street sky view factor; and f is the binary indicator of obstacles. Ψ_{sky} and f can be estimated using GSV images.

Clear-Sky Street-Level Solar Irradiance

The value of clear-sky solar irradiance primarily depends on the morphologies and orientations of a street canyon, and, therefore, has direct implications on urban design at street level. Based on calculations described in Jendritzky (1990) and

3.4 GSV-Based Estimation of Street-Level Solar Radiation

Matzarakis et al. (2010), the direct irradiance I_{open} on a horizontal surface in a free horizon with no obstructions can be calculated by:

$$I_{open} = E_0 \times \cos\varphi \times \exp\left(-T_L \times \delta_{r_0} \times \frac{p}{p_0} \times m_{r_0}\right) \quad (3.9)$$

where E_0 is the extraterrestrial radiation; T_L is the Linke turbidity factor; δ_{r_0} is the vertical optical thickness of the standard atmosphere; and m_{r_0} is the relative optical air mass, which considers the extended optical path through the atmosphere with respect to vertical incidence. The relative optical air mass is calculated (Kasten & Young, 1989) by:

$$m_{r_0} = 1/\left(\sin\gamma + 0.50572 \times (\gamma + 6.07995°)^{-1.6364}\right) \quad (3.10)$$

where the solar altitude angle $\gamma = 90° - \varphi$. The optical thickness can be calculated (Kasten, 1996) by

$$\delta_{r_0} = 1/0.9m_{r_0} + 9.4 \quad (3.11)$$

for zenith angle $\varphi < 85°$, i.e. $\gamma > 5°$.

According to the calculations as described in Valko (1966) and Matzarakis et al. (2010), the calculations of D_{iso_open}, D_{aniso_open}, and D_{cloud_open} are shown below. N is cloudiness in the unit of octas. Under the cloudless assumption for calculating clear-sky solar irradiance, $N = 0$.

$$D_{iso\_open} = (G_0 - I_{open}) \times (1-\tau) \times (1 - N/8) \quad (3.12)$$

$$D_{aniso\_open} = (G_0 - I_{open}) \times \tau \times (1 - N/8) \quad (3.13)$$

$$D_{cloud\_open} = 0.28 \times G_0 \times N/8 \quad (3.14)$$

where $\tau = I/(E_0 \cos\varphi)$ is the transmittance of the direct solar radiation.

The global irradiance G_0 for undisturbed conditions (free horizon and no clouds) can be estimated (Jendritzky, 1990) as follows:

$$G_0 = 0.84 \times E_0 \times \cos\varphi \times \exp\left(-0.027 \times \frac{p}{p_0} \times T_L / \cos\varphi\right) \quad (3.15)$$

The variables in Eq. (3.15) have the same meaning as Eq. (3.9), with the solar radiation flux density E_0 (W/m^2), the zenith angle φ (°) of the sun, the local atmospheric pressure p (hPa) relative to the normal pressure $p_0 = 1013$ hPa at sea level, and the Linke turbidity factor T_L in Hong Kong (Li & Lam, 2002). As an example, Fig. 3.8 illustrates the calculation results of solar hours and clear-sky solar irradiation in a street canyon. However, for applications in urban areas, it is necessary to account for cloudiness, as described by the all-sky street-level solar irradiance.

All-Sky Street-Level Solar Irradiance

In this study, the all-sky solar irradiance measurements without horizontal obstructions in KP from HKO are used as inputs in Eqs. (3.7) and (3.8) to calculate the all-sky street-level solar irradiance. Our use of solar radiation in KP for the whole study area is investigated in the discussion section in which we found that the free-horizon solar radiation in Hong Kong can be assumed to be homogeneous since the difference between KP (urban site) and KSC (rural site) is relatively small.

The direct irradiance measurement is the I_{open} in Eq. (3.7). To decompose the diffuse radiation measurements to get the D_{iso_open}, D_{aniso_open}, and D_{cloud_open} in Eq. (3.8), we calculate these three components using Eqs. (3.12), (3.13), and (3.14) and cloudiness measurements from HKO. Calculations of these components are then scaled to fit the diffuse radiation measurements assuming the fractions from the calculations are relatively accurate. Eventually, all-sky solar irradiance of street canyons can be calculated using Eqs. (3.6), (3.7), and (3.8). The results of all-sky global irradiation of street canyons and its direct and diffuse components are shown in Sects. 5.3.2 and 5.3.3, respectively.

Isotropic and Anisotropic Components of Diffuse Radiation

The diffuse radiation is the combination of isotropic and anisotropic components. When the diffuse radiation transfers to the street level, as shown in Eq. (3.8), isotropic (D_{iso}) component is associated with the value of Ψ_{sky}, while anisotropic (D_{aniso}) component is affected by the urban geometry of street canyons and f, the binary indicator of the existence of solar ray path obstructions.

Figure 3.9 shows the direct radiation, and isotropic and anisotropic components of diffuse radiation under free-horizon condition. B_1 in black is the isotropic diffuse component, which is mainly caused by the Rayleigh scattering from gas molecule; B_2 in blue is the anisotropic diffuse component, which is mainly caused by Mie scattering from atmospheric aerosols. Mie scattering has a much stronger scattering around the forward scattering direction which overlaps with direct solar radiation. The details are shown in Fig. 3.9b. When there is large particulate matter in the air (the air pollutants in Hong Kong), the forward part of Mie scattering is dominant. Since it does not have a strong wavelength dependence, we see a white glare around the sun. Since Rayleigh scattering strongly favors short wavelengths, we see an isotropic blue sky.

To compare the relative magnitude of isotropic and anisotropic diffuse components, we calculate them at free-horizon at different solar zenith angles, as shown in Fig. 3.10. In general, both isotropic and anisotropic components decrease as solar zenith angle increases due to the decrease of the vertical component which

3.4 GSV-Based Estimation of Street-Level Solar Radiation

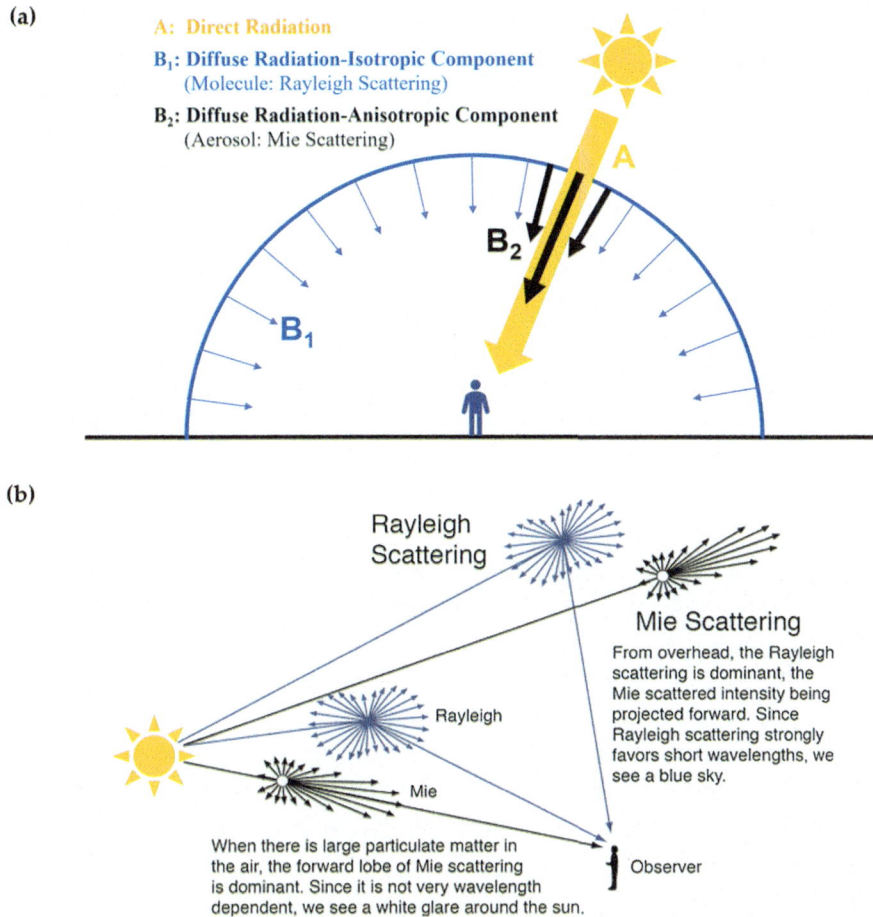

Fig. 3.9 (a) The direct radiation, and isotropic and anisotropic components of the diffuse radiation incident on the street surface. (b) The Rayleigh scattering from atmospheric molecules and Mie scattering from large atmospheric particles

is perpendicular to the surface. The isotropic component, however, reaches the peak at 60°, when the increase of air mass and a decrease of vertical component balances each other. The anisotropic component is larger than isotropic when the SZA is small. However, when SZA is larger than 35°, the isotropic component is higher. We can see the total diffuse radiation, the sum of the two components, is highly anisotropic, indicating that an anisotropic sky exists for all SZAs in Hong Kong.

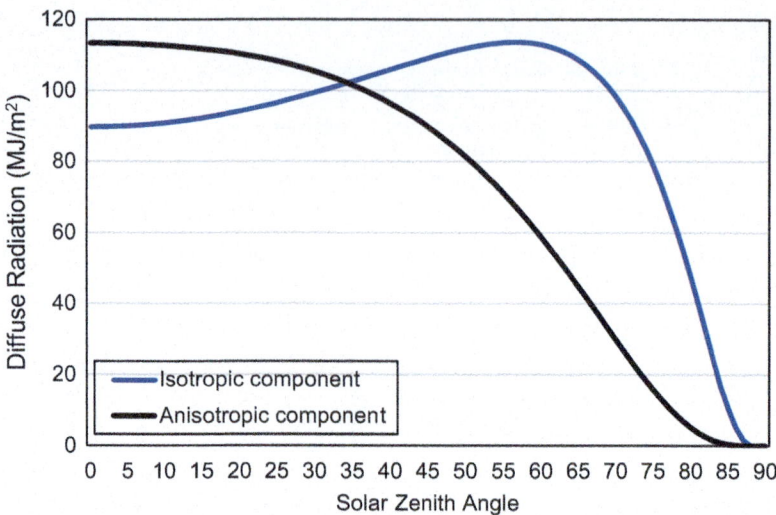

Fig. 3.10 The change of isotropic and anisotropic components of diffuse radiation with solar zenith angle from 0 to 90 under free-horizon condition

3.5 Summary

This chapter illustrates the methodology of Google Street View (GSV)-based street view factors (Part I) and solar radiation estimations (Part II). This study takes Hong Kong, a typical high-density urban context as the experimental area. The methodological framework mainly made up two board Parts:

- Part I focuses on developing an approach for accurately deriving view factors for the sky, trees, and buildings of street canyons in the high-density urban environment of Hong Kong using the publicly available GSV images and a deep-learning feature extraction algorithm;
- Based on the results of Part I, Part II further focuses on developing an approach for accurately calculating the street-level global, direct, and diffuse solar irradiance in high-density urban environments using publicly available GSV images;
- The developed method of Part I for analyzing VFs in a 3-D street environment will play an important role in relating science-based evidence for urban climatic studies and decision-making in urban planning and design processes. The developed method of Part II for mapping the street-level solar radiation can be, on a global scale, applied to cities with available coverage of GSV images. This method provides a low-cost and effective streetscape mapping approach for urban studies. Clear-sky street-level solar irradiance can be derived for cities with basic meteorological measurements including atmospheric pressure while all-sky street-level solar irradiance can be further calculated if free-horizon solar irradiance is available at a local site, such as HKO measurements in Hong Kong.

The further results of the spatial patterns of the sky, tree, and building view factor of the study area will be analyzed in Chap. 4; The further results of the spatial pattern and temporal variation of street-level solar irradiance of the study area will be analyzed in Chap. 5.

References

Anguelov, D., Dulong, C., Filip, D., Frueh, C., Lafon, S., Lyon, R., & Weaver, J. (2010). Google street view: Capturing the world at street level. *Computer, 43*(6), 32–38. https://doi.org/10.1109/MC.2010.170

Badrinarayanan, V., Kendall, A., & Cipolla, R. (2017). Segnet: A deep convolutional encoder-decoder architecture for image segmentation. *IEEE Transactions on Pattern Analysis and Machine Intelligence, 39*(12), 2481–2495. https://doi.org/10.1109/TPAMI.2016.2644615

Census and Statistics Department. (2016). *2011 Hong Kong population census*. Retrieved October 31, 2018, from https://www.census2011.gov.hk/en/index.html

Census and Statistics Department, The Government of Hong Kong S A R. (2016). *Population - Overview | Census and statistics department*. Retrieved October 25, 2017, from http://www.censtatd.gov.hk/hkstat/sub/so20.jsp

Erell, E., Pearlmutter, D., Boneh, D., & Kutiel, P. B. (2014). Effect of high-albedo materials on pedestrian heat stress in urban street canyons. *Urban Climate, 10*, 367–386. https://doi.org/10.1016/j.uclim.2013.10.005

Gong, F.-Y. (2019). *Mapping street canyon morphology and solar radiation in high-density urban environments using street sensing approach*. The Chinese University of Hong Kong.

Gong, F.-Y., Zeng, Z.-C., Ng, E., & Norford, L. K. (2019). Spatiotemporal patterns of street-level solar radiation estimated using Google street view in a high-density urban environment. *Building and Environment, 148*, 547–566. https://doi.org/10.1016/j.buildenv.2018.10.025

Gong, F.-Y., Zeng, Z.-C., Zhang, F., Li, X., Ng, E., & Norford, L. K. (2018). Mapping sky, tree, and building view factors of street canyons in a high-density urban environment. *Building and Environment, 134*, 155–167. https://doi.org/10.1016/j.buildenv.2018.02.042

Google Maps APIs. (2017a). *Google street view image API | Google Street View Image API*. Retrieved October 20, 2017, from https://developers.google.com/maps/documentation/streetview/intro

Google Maps APIs. (2017b). *Street view image metadata | Google Street View Image API*. Retrieved November 27, 2017, from https://developers.google.com/maps/documentation/streetview/metadata

Google Maps APIs. (2017c). *Street view service | Google Maps JavaScript API*. Retrieved March 21, 2018, from https://developers.google.com/maps/documentation/javascript/streetview

Google Street View. (2017). *Google street view—Where we've been & where we're headed next*. Retrieved January 18, 2018, from https://www.google.com/streetview/understand/

Hong Kong Observatory. (2003a). *24-hour time series of mean sea level pressure in Hong Kong*. Retrieved May 13, 2018, from http://www.hko.gov.hk/wxinfo/ts/display_element_pp_e.htm

Hong Kong Observatory. (2003b). *24-hour time series of solar radiation*. Retrieved May 12, 2018, from http://www.hko.gov.hk/wxinfo/ts/display_element_solar_e.htm

Hong Kong Observatory. (2003c). *King's park meteorological station*. Retrieved March 21, 2018, from http://www.hko.gov.hk/wxinfo/aws/kpinfo.htm

Hong Kong Observatory. (2010). *Climate of Hong Kong*. Retrieved June 22, 2018, from http://www.weather.gov.hk/cis/climahk_e.htm

Hong Kong Observatory. (2012). *Direct and diffuse solar radiation information added to observatory's website*. Retrieved February 22, 2018, from http://www.hko.gov.hk/press/D4/pre20100401e.htm

Hong Kong Observatory. (2016). *The year's weather - 2016*. Retrieved November 27, 2017, from http://www.hko.gov.hk/wxinfo/pastwx/2016/ywx2016.htm

Hong Kong Observatory. (2018). *Climate change in Hong Kong - Cloud amount, solar radiation and evaporation*. Retrieved May 24, 2018.

Jendritzky, G. (1990). Methodik zur räumlichen Bewertung der thermischen Komponente im Bioklima des Menschen: fortgeschriebenes Klima-Michel-Modell. na.

Johnson, G. T., & Watson, I. D. (1984). The determination of view-factors in urban canyons. *Journal of Climate and Applied Meteorology, 23*(2), 329–335. https://doi.org/10.1175/1520-0450(1984)023<0329:TDOVFI>2.0.CO;2

Kasten, F. (1996). The linke turbidity factor based on improved values of the integral Rayleigh optical thickness. *Solar Energy, 56*(3), 239–244. https://doi.org/10.1016/0038-092X(95)00114-7

Kasten, F., & Young, A. T. (1989). Revised optical air mass tables and approximation formula. *Applied Optics, 28*(22), 4735. https://doi.org/10.1364/AO.28.004735

LeCun, Y., Bengio, Y., & Hinton, G. (2015). Deep learning. *Nature, 521*(7553), 14539. https://doi.org/10.1038/nature14539

Li, D. H. W., & Lam, J. C. (2002). A study of atmospheric turbidity for Hong Kong. *Renewable Energy, 25*(1), 1–13. https://doi.org/10.1016/S0960-1481(01)00008-8

Li, X., Ratti, C., & Seiferling, I. (2018). Quantifying the shade provision of street trees in urban landscape: A case study in Boston, USA, using Google street view. *Landscape and Urban Planning, 169*(Supplement C), 81–91. https://doi.org/10.1016/j.landurbplan.2017.08.011

Lin, T.-Y., Dollár, P., Girshick, R., He, K., Hariharan, B., & Belongie, S. (2016). *Feature pyramid networks for object detection.* ArXiv:1612.03144 [Cs]. Retrieved from http://arxiv.org/abs/1612.03144

Long, J., Shelhamer, E., & Darrell, T. (2015). *Fully convolutional networks for semantic segmentation* (pp. 3431–3440). IEEE. Retrieved from https://www.cv-foundation.org/openaccess/content_cvpr_2015/html/Long_Fully_Convolutional_Networks_2015_CVPR_paper.html

Lu, W.-Z., He, H., & Leung, A. Y. T. (2011). Assessing air quality in Hong Kong: A proposed, revised air pollution index (API). *Building and Environment, 46*(12), 2562–2569. https://doi.org/10.1016/j.buildenv.2011.06.011

Matzarakis, A., Rutz, F., & Mayer, H. (2007). Modelling radiation fluxes in simple and complex environments—Application of the RayMan model. *International Journal of Biometeorology, 51*(4), 323–334. https://doi.org/10.1007/s00484-006-0061-8

Matzarakis, A., Rutz, F., & Mayer, H. (2010). Modelling radiation fluxes in simple and complex environments: Basics of the RayMan model. *International Journal of Biometeorology, 54*(2), 131–139. https://doi.org/10.1007/s00484-009-0261-0

National Renewable Energy Laboratory. (2018). *MIDC: Solar position and intensity (SOLPOS) calculator.* Retrieved March 29, 2018, from https://midcdmz.nrel.gov/solpos/solpos.html

Ng, E., & Cheng, V. (2012). Urban human thermal comfort in hot and humid Hong Kong. *Energy and Buildings, 55*(Supplement C), 51–65. https://doi.org/10.1016/j.enbuild.2011.09.025

Planning Department, The Government of the Hong Kong, S A R. (2016). *Planning department - Land utilization in Hong Kong.* Retrieved October 26, 2017, from http://www.pland.gov.hk/pland_en/info_serv/statistic/landu.html

Stewart, I. D., & Oke, T. R. (2012). Local climate zones for urban temperature studies. *Bulletin of the American Meteorological Society, 93*(12), 1879–1900. https://doi.org/10.1175/BAMS-D-11-00019.1

Valko, P. (1966). Die Himmelsstrahlung in ihrer Beziehung zu verschiedenen Parametern. *Archiv für Meteorologie, Geophysik und Bioklimatologie Serie B, 14*(3–4), 336–359. https://doi.org/10.1007/BF02243366

Wang, W., Zhou, W., Ng, E. Y. Y., & Xu, Y. (2016). Urban heat islands in Hong Kong: Statistical modeling and trend detection. *Natural Hazards, 83*(2), 885–907. https://doi.org/10.1007/s11069-016-2353-6

Watson, I. D., & Johnson, G. T. (1987). Graphical estimation of sky view-factors in urban environments. *Journal of Climatology, 7*(2), 193–197. https://doi.org/10.1002/joc.3370070210

Yang, Z., & Gong, F.-Y. (2025). Utilizing street view images to estimate solar energy potential for photovoltaic-powered buses. *Applied Geography, 177*, 103567. https://doi.org/10.1016/j.apgeog.2025.103567

Zhao, H., Shi, J., Qi, X., Wang, X., & Jia, J. (2016). *Pyramid scene parsing network.* ArXiv:1612.01105 [Cs]. Retrieved from http://arxiv.org/abs/1612.01105

Zheng, Y., Ren, C., Xu, Y., Wang, R., Ho, J., Lau, K., & Ng, E. (2017). GIS-based mapping of local climate zone in the high-density city of Hong Kong. *Urban Climate.* https://doi.org/10.1016/j.uclim.2017.05.008

Zhou, B., Zhao, H., Puig, X., Fidler, S., Barriuso, A., & Torralba, A. (2016). *Semantic understanding of scenes through the ADE20K dataset.* ArXiv:1608.05442 [Cs]. Retrieved from http://arxiv.org/abs/1608.05442

Chapter 4
Spatial Patterns of Street Canyon View Factors

Contents

4.1	Overview.	52
4.2	Methods Verification.	52
	4.2.1 Accuracy Assessment of Features Classification by Deep Learning.	52
	4.2.2 Verification of GSV-Based View Factor Estimates.	54
	4.2.3 3D-GIS-Based SVF Estimates.	55
4.3	Results.	57
	4.3.1 Mapping SVF, TVF, and BVF of Street Canyons Using GSV Images.	57
	4.3.2 Comparison between GSV-Based and 3D-GIS-Based SVF Estimates.	59
	4.3.3 Impacts of Street Tree Canopy and Building Density on SVF Estimates.	59
4.4	Discussion.	62
	4.4.1 Large Uncertainty in Model-Based SVF Estimates from Street Trees.	62
	4.4.2 Temporal Variation of Street-Level View Factors.	63
	4.4.3 Spatial Variation of Street-Level View Factors.	65
4.5	Summary.	65
References.		66

Abstract This chapter presents a novel approach to quantify sky view factor (SVF), tree view factor (TVF), and building view factor (BVF) in Hong Kong's high-density urban environment using Google Street View (GSV) imagery and a deep-learning-based scene parsing algorithm. The method addresses the complexity of urban geometry, including building overhangs and tree canopy cover, by analyzing 29,264 GSV images at 30-m intervals. Validation against hemispheric photography field measurements confirms high accuracy ($R^2 > 0.95$ for SVF, TVF, and BVF), marking the first direct verification of GSV-derived view factors using fisheye lens data. A comparative analysis with conventional 3D-GIS modeling reveals significant overestimation of SVF by the 3D-GIS method (average difference: 0.11), attributed to its inability to account for street tree obstruction. Spatial mapping demonstrates that high-density areas exhibit lower SVF (0.49 average) and TVF (0.14 average), with tree coverage strongly correlated to discrepancies between GSV and 3D-GIS results

($R^2 = 0.53$). The study highlights the superiority of GSV-based methods in capturing ground-level urban complexity and underscores the necessity of integrating street tree data into urban radiative models for improved microclimate assessments.

Keywords Sky view factor (SVF) · Google street view (GSV) · Deep learning · Urban geometry · High-density cities

4.1 Overview

View factor is a geometric ratio that expresses the fraction of the radiation output from one surface that is intercepted by another (Oke, 1987). View factors for the sky, trees, and buildings are three important parameters of the urban outdoor environment that describe the geometrical relation between different surfaces from the perspective of radiative energy transfer. Gong et al. (2018) develops an approach for accurately estimating sky view factor (SVF), tree view factor (TVF), and building view factor (BVF) of street canyons in the high-density urban environment of Hong Kong using publicly available Google Street View (GSV) images and a deep-learning algorithm for extraction of street features (sky, trees, and buildings).

The purpose of this chapter is to develop an approach for estimating and mapping SVF, TVF, and BVF of street canyons with complex urban living environment context, like the high-density urban areas of Hong Kong. The approach is based on GSV images and a deep-learning technique for street feature extraction and is validated using hemispheric photography measurements as reference data from fieldwork. This verification is, to our knowledge, the first reported use of hemispheric photography for direct verification of GSV-based streetscape study. The developed approach represents a ground-based perspective of city streetscapes that cover complicated urban contexts, including tree canopy cover, building overhangs, and shade structures. A comparison with conventional 3-D modeling of SVF, which was widely used in previous studies, is also conducted to assess the uncertainty and advantages of this GSV-based mapping approach.

4.2 Methods Verification

4.2.1 Accuracy Assessment of Features Classification by Deep Learning

To assess the accuracy of the scene parsing deep-learning technique in extracting street scenes, especially for the VFs of the sky, trees, and buildings, the focus of this study, we randomly select 100 sampled street points and collect their corresponding

4.2 Methods Verification

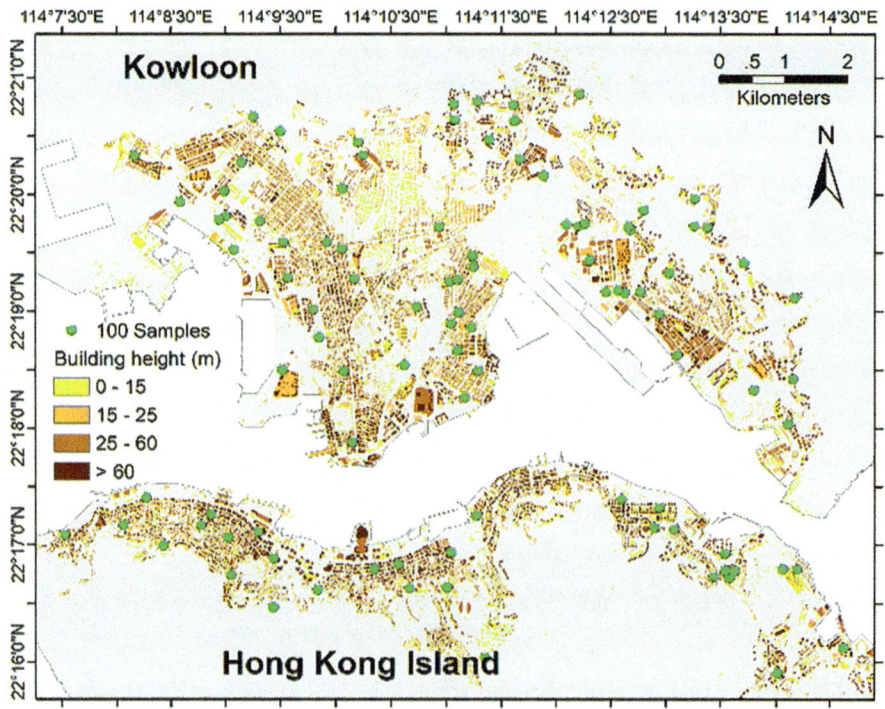

Fig. 4.1 Spatial distribution of the 100 randomly selected samples in the study area for verification

GSV images. Figure 4.1 shows the spatial distribution of the 100 randomly selected samples. The sampling is implemented by generating 100 random integers using a pseudorandom integer generator in MATLAB to extract data from the total 29,264 street data points. The samples are randomly distributed in the study area and covers from low building density to high building density (SVF values vary from 0.0 to 1.0 as shown in Fig. 4.2), indicating these samplings are representative of different characteristics of street canyons and street trees. A manual delineation on the images by eye inspection is implemented to extract the sky feature to generate a reference dataset (as truth). As a result, Fig. 4.2 shows the comparison of calculated VFs from GSV images based on deep-learning technique and the generated reference data. The two datasets exactly agree with each other, with R^2 as high as 0.974 and RMSE of 0.036 for SVF, R^2 of 0.986 and RMSE of 0.025 for TVF, and R^2 of 0.983 and RMSE of 0.037 for BVF. This agreement suggests that the scene parsing deep-learning technique is able to accurately extract the street-level features in high-density urban areas of Hong Kong.

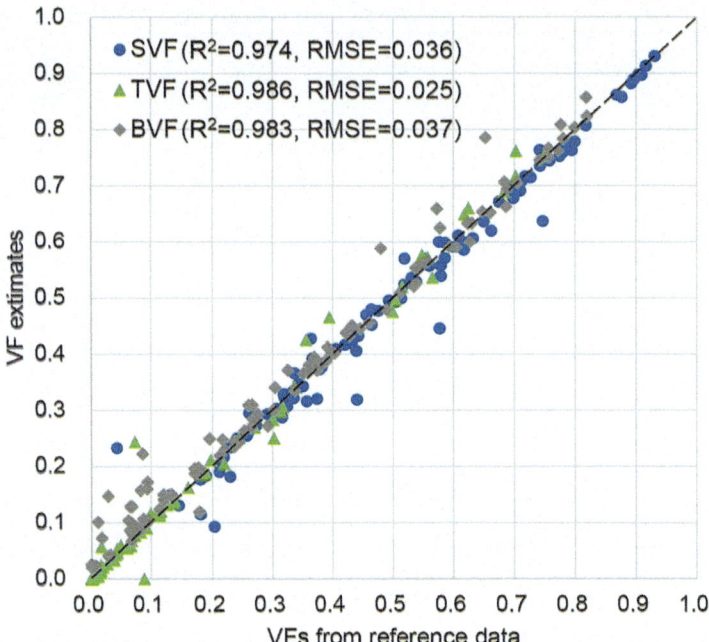

Fig. 4.2 Accuracy assessment of feature extraction using the PSPNet in a deep-learning framework to calculate SVF, TVF, and BVF from GSV images in high-density urban areas of Hong Kong. The R^2 and RMSE between the two datasets of VFs are also indicated

4.2.2 Verification of GSV-Based View Factor Estimates

Comparison of VF estimates and direct measurements using hemispherical photography is a convincing way to verify the effectiveness and assess the uncertainties associated with GSV-based and 3D-GIS-based methods in estimating VFs of street canyons. Here we use the fisheye lens hemispheric photography to verify the applicability of GSV-based and 3D-GIS-based methods in high-density urban areas of Hong Kong. This is, to our knowledge, the first reported use of hemispheric photography for direct verification of a GSV-based streetscape study.

Forty photographs were taken at 40 selected sample points (20 in the high-rise area of Mong Kok and 20 in the low-rise area of Kowloon Tong) in Kowloon, as shown in the dotted red and blue rectangles, respectively, in Fig. 3.1c. The photographs were taken at 1.5 m above ground level, using a digital camera, Nikon FM601, with an 8-mm circular lens. Figure 4.3 shows four examples of photographs taken with a fisheye lens in our field survey (a) and the collocated projected GSV fisheye images (b). The feature extraction results and the corresponding VF estimates are also shown. The field survey results are consistent with GSV-based estimates, with differences within 0.03, suggesting the effectiveness and high accuracy of using GSV images in estimating VFs in high-density urban areas of Hong Kong.

Figure 4.4 illustrates the comparisons of survey-based reference VF data and GSV-based and 3D-GIS-based VF estimates. The GSV-based method we propose to

4.2 Methods Verification

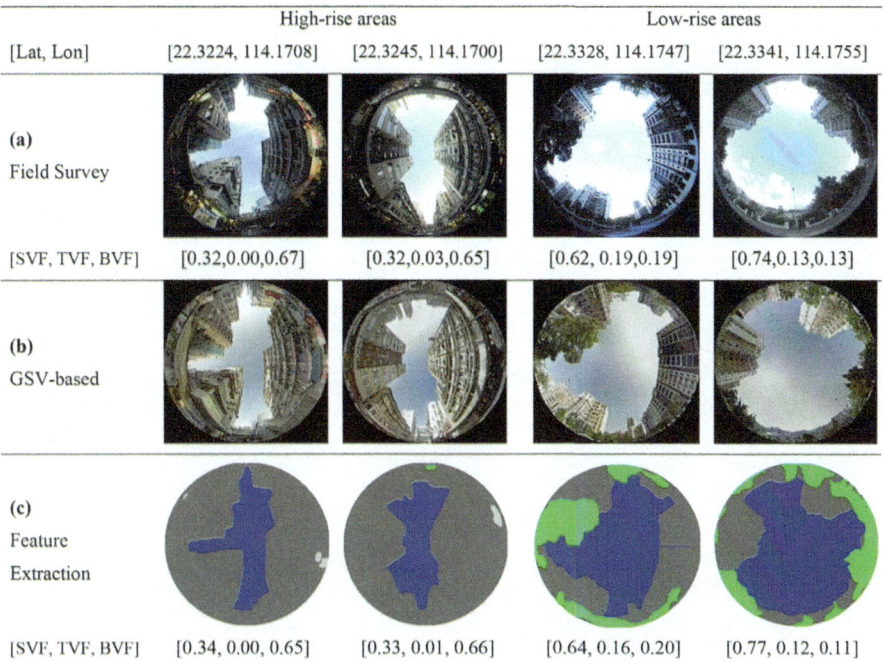

Fig. 4.3 Examples of fisheye images from two high-rise and two low-rise street sample points from a field survey in (**a**), and GSV-based method in (**b**). Image features are classified into the sky (in blue), tree (in green), and building (in gray) using the scene parsing deep-learning technique, as shown in (**c**). SVF, TVF, and BVF values from field surveys and GSV are shown as indicated

use in high-density urban areas of Hong Kong performs much better in estimating SVF (Fig. 4.4a) than the commonly used 3D-GIS-based method, with higher R^2 (0.954 versus 0.014) and lower RMSE (0.033 versus 0.263). In particular, 3D-GIS has higher R^2 in high-rise areas than in low-rise areas, indicating that model simulation performs better in high-rise areas. Moreover, for GSV-based estimates, R^2 and RMSE for TVF (Fig. 4.4b) are 0.987 and 0.027, respectively, and for BVF (Fig. 4.4c) they are 0.986 and 0.036, respectively. These results suggest that a GSV-based method is effective and accurate in high-density urban areas of Hong Kong, characterized by compact high-rise areas with complicated street environments and by low-rise areas with dense tree canopy.

4.2.3 3D-GIS-Based SVF Estimates

The high-resolution 3D-GIS-based SVF estimates of Hong Kong was developed by Chen et al. (2012) in studying Hong Kong's urban microclimate. The 3D-GIS Model is generated in the geographic information system (GIS) by using a 3-D building database (with building height information) merged with the topography database to create a digital elevation layer representing the height of the urban

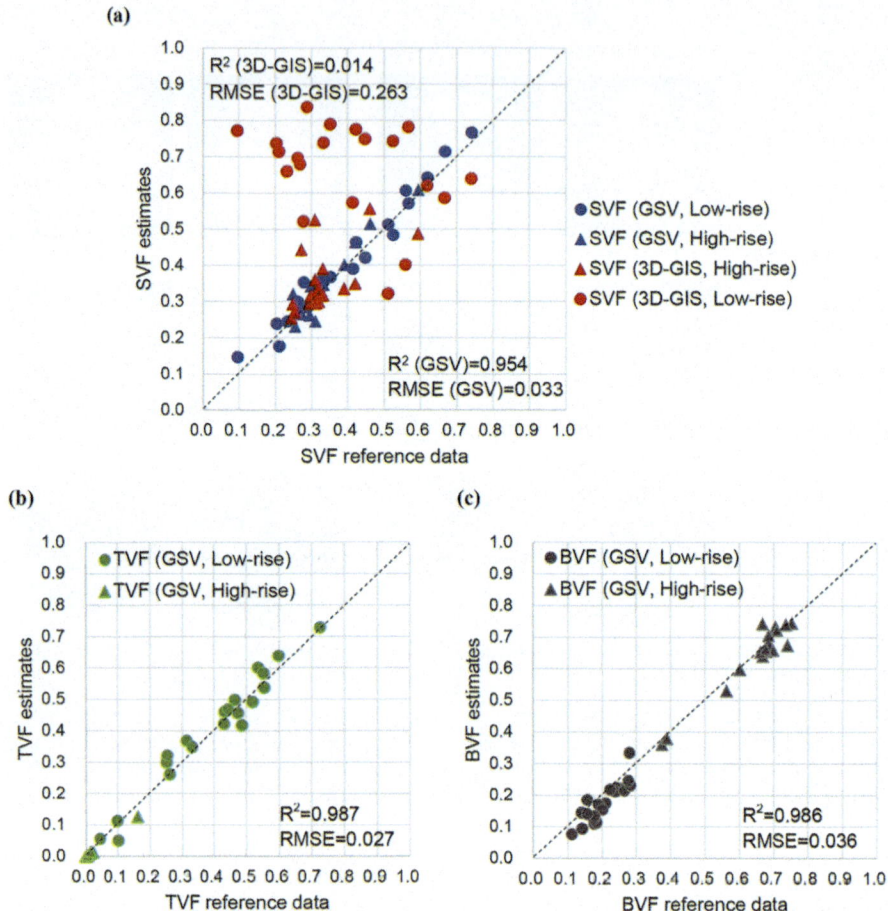

Fig. 4.4 (a) Scatter plot of SVF reference data from field survey and the corresponding GSV-based (in blue) and 3D-GIS-based (in red) SVF estimates. Sampling SVF data include 20 samples in Mong Kok within high-rise building area (in triangles), and 20 samples in Kowloon Tong within the low-rise area (in circles); (b) the same as (a) but for TVF; (c) the same as (a) but for BVF

surface in Hong Kong. Both 3-D building and topography databases are provided by the Hong Kong Planning Department. Continuous SVF values at 2-m resolution are calculated for an entire urban environment. The SVF is calculated by first constructing a fisheye image and then estimating the SVF value using Eq. (3) in Johnson and Watson (1984), the same way with using GSV images as described in Sect. 3.3.3.

In this study, SVF estimates from the 3D-GIS model are compared with those from GSV, as described in Sect. 4.3.2, for the purpose of assessing the accuracy of urban 3D-GIS model in simulating urban street environments under high-density contexts This comparison is important because: (1) 3D-GIS model has been widely used to estimate the street geometric structures, including SVF, but a comprehensive verification in a wide region is not possible because the conventional verification method is both time and effort consuming (Chen et al., 2012); (2) GSV-based method makes it possible to map the SVF at street level across the whole city, and therefore, as this study shows, a comprehensive verification of the 3D-GIS model can be conducted. The understanding of the discrepancies between GSV-based and 3D-GIS-based SVF estimates sheds light on future improvement in the model simulation of the complex urban environments.

4.3 Results

4.3.1 Mapping SVF, TVF, and BVF of Street Canyons Using GSV Images

Figure 4.5 shows the spatial distributions of GSV-based SVF, TVF, and BVF estimates in high-density urban areas in Kowloon and Hong Kong Island, and a comparison of their frequency distributions. The average SVF, TVF, and BVF values in high-density areas of Hong Kong are 0.49, 0.14, and 0.33, respectively, and there are small differences between Kowloon area (0.53, 0.12, and 0.41) and Hong Kong Island (0.41, 0.19, and 0.36).

The SVF value ranges from near 0, indicating little sky openness, to 1.0, indicating total sky openness. In general, we found the spatial patterns of VF estimates are similar and consistent with the corresponding building height and density (see Fig. 3.1c). Areas with higher density have lower SVF, lower TVF, and higher BVF, and vice versa. The high-density residential areas, located in southern and western Kowloon and northern Hong Kong Island, which cover about 58% of the study area, are dominated by low and median SVF (0.2–0.6), and low TVF (0.0–0.3), because of the high-density construction and narrow streets that block sky visibility and limit space for greenery. The coastline regions and low-rise areas, which cover about 20% of the study area, show much higher SVF (0.7–1.0), and lower BVF (0.0–0.3), because of fewer buildings and more sky openness.

In low-rise regions near country parks in the southern part of the study area in Hong Kong Island where BVF is low (e.g., the dotted rectangles in Fig. 4.5), the SVF values, however, are in a much lower range (0.0–0.2; Fig. 4.5c). The much lower sky openness, as we discuss later, is mainly due to high tree cover in this area (Fig. 4.5b), which blocks much of the sky visibility. Further analysis of the impact due to tree cover is described in Sect. 4.3.3. Figure 4.5d shows the frequency distribution of the VFs in Hong Kong. The TVF in the high building density area is dominated by values less than 0.1 and the average value is 0.143. The low TVF is mainly limited by the high building density and narrow streets. This average TVF is smaller compared with Singapore (0.293), a subtropical high building and population density city in Asia, and is similar to New York (0.135), a typical high-density city in the United States (MIT Senseable City Lab, 2016). SVF, on the other hand, is close to an even distribution between 0.2 and 0.9, with a peak between 0.4 and 0.5, while BVF has a decreasing frequency when its value increases.

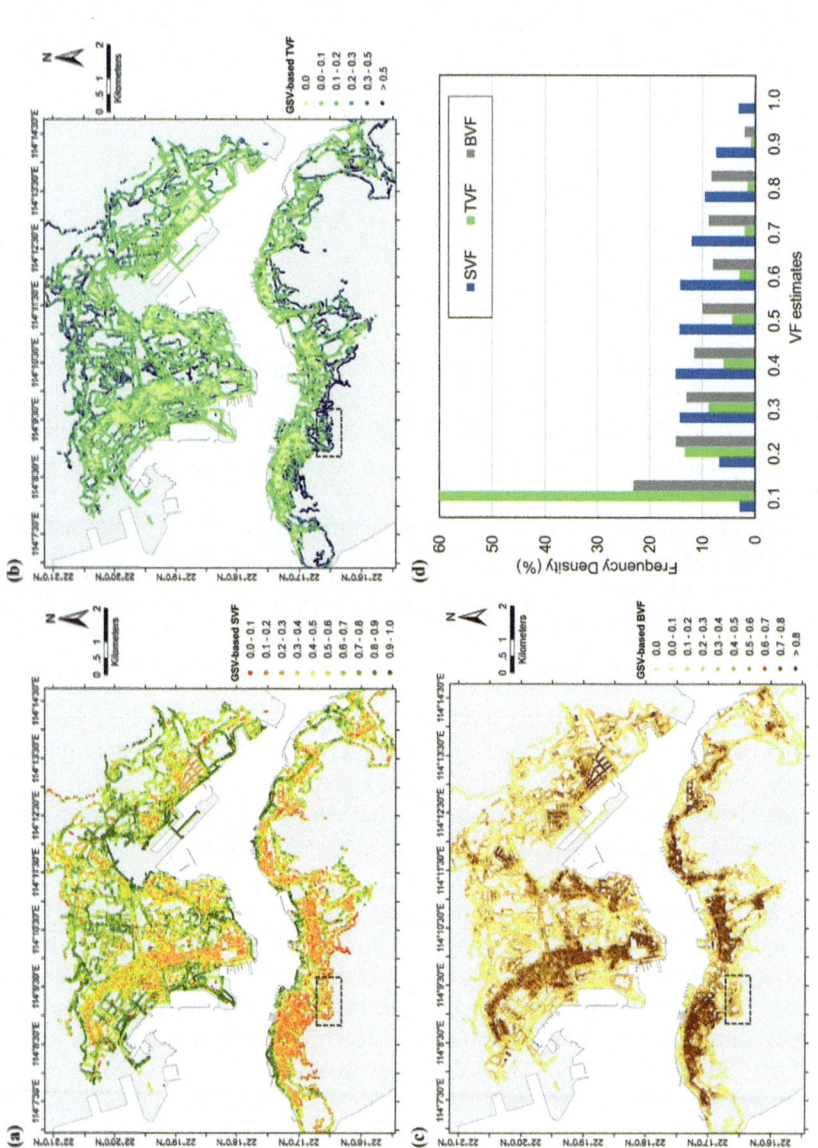

Fig. 4.5 Maps of GSV-based SVF in (a), TVF in (b), and BVF in (c) of street canyons in high-density urban areas of Hong Kong derived from 29,264 GSV images along streets at 30-m intervals; (d) Frequency density of SVF (blue bar), TVF (green bar), and BVF (gray bar)

4.3 Results

4.3.2 Comparison between GSV-Based and 3D-GIS-Based SVF Estimates

In this section, the collocated SVF estimates at 30-m intervals derived from GSV and 3D-GIS Model are compared for the purpose of assessing the accuracy of urban 3D-GIS model in simulating urban street environment under high-density contexts. Figure 4.6a illustrates the spatial distribution of 3D-GIS-based SVF estimates at 30-m intervals in high-density areas of Hong Kong, corresponding to the sampling points shown in Fig. 4.5a. The map of 3D-GIS-based SVF shows a similar pattern to that of GSV-based SVF estimates, in which the lower SVF values are located mainly in areas with high-rise buildings and higher SVF values located in low-rise or coastal areas. The average SVF value of 3D-GIS-based estimates (0.59) is about 0.11 (about 20%) higher than that of GSV-based estimates (0.49) but with the similar standard deviation (0.234 and 0.225, respectively). However, there are large differences in the low-rise areas with large quantities of street trees. To investigate the spatial difference, Fig. 4.6b shows the spatial distribution of the difference between 3D-GIS-based and GSV-based SVF estimates (former minus latter). The two datasets have a better agreement in high-rise regions (difference less than 0.1) than the regions with lower building rise (difference larger than 0.1). This characteristic of difference can also be seen from the bivariate histogram in Fig. 4.6c, which shows that (1) GSV-based and 3D-GIS-based SVF estimates have good agreement (the highest data number density can be seen in the diagonal direction) in high-rise regions with small SVF between 0.2 and 0.4. Out of all the sampling points, 43.85% of them have a difference larger than 0.1; (2) some 3D-GIS-based values are higher than GSV-based values, suggesting model simulations overestimate the SVF in some regions. According to the descriptive statistics of the 29,264 sample points from this comparison, the R^2 between them is 0.40 with RMSE of 0.22. This overestimation of SVF by the 3D-GIS-based method can also be seen in the histogram plots in Fig. 4.6d, which shows that the 3D-GIS-based method has shifted the peak of the frequency distribution of SVF from less than 0.5 to larger than 0.5. As a result, the averaged SVFs are different by 0.11. The contributors to this pattern of difference are investigated in Sect. 4.3.3.

4.3.3 Impacts of Street Tree Canopy and Building Density on SVF Estimates

To gain further understanding of the discrepancies between GSV-based and 3D-GIS-based SVF estimates and shed light on future improvements in the model simulation of the urban environment, we investigate the impacts of the street tree canopy, quantified using the TVF estimate, and building density, quantified using the BVF estimate, on the discrepancies are shown in Fig. 4.6.

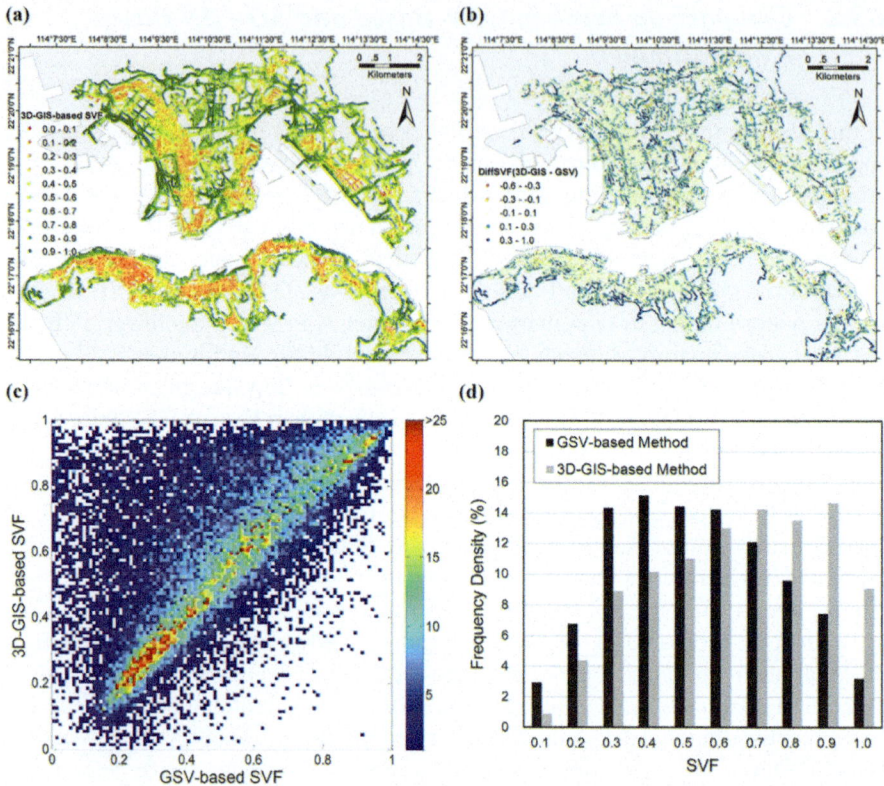

Fig. 4.6 (a) Map of 3D-GIS-based SVF estimate with the same street sampling points as shown in Fig. 4.5a; (b) Map of the difference between GSV-based and 3D-GIS-based SVF estimates; (c) Bivariate histogram of GSV-based and 3D-GIS-based SVF estimates of street canyon in high-density urban areas of Hong Kong as shown in Figs. 4.5a and 4.6a, respectively. To make the histogram, the SVF data from both datasets are grouped into 0.01 × 0.01 grids and the value of a grid is the total number of SVF samples that fall in this grid; (d) Comparison of frequency density histogram from GSV-based and 3D-GIS-based SVF estimates

The bivariate histogram of GSV-based TVF and the difference between GSV-based and 3D-GIS-based SVF estimates are presented in Fig. 4.7a. We can see that (1) a large majority of the data has small differences (close to 0) and low TVF (smaller than 0.2) and (2) there is a strong positive correlation between TVF and the difference. When TVF is larger than 0.1, R^2 of the two datasets is 0.53 ($p < 0.01$), indicating a significant correlation. This result indicates that a higher number of street trees leads to a larger difference between estimations of SVF from GSV images and 3D-GIS simulation, especially when the TVF is larger than 0.1. The result from linear regression shows that the increase of the difference follows the increase of TVF by a factor of 1.17. This strong correlation suggests that TVF, an indicator of the amount of street trees, makes the dominant contribution to the discrepancies between GSV-based and 3D-GIS-based SVF estimates. This is because

4.3 Results

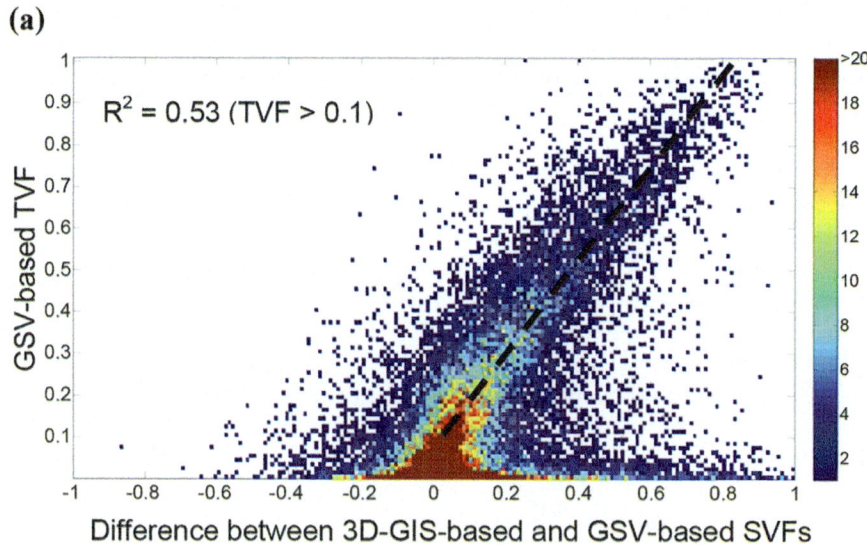

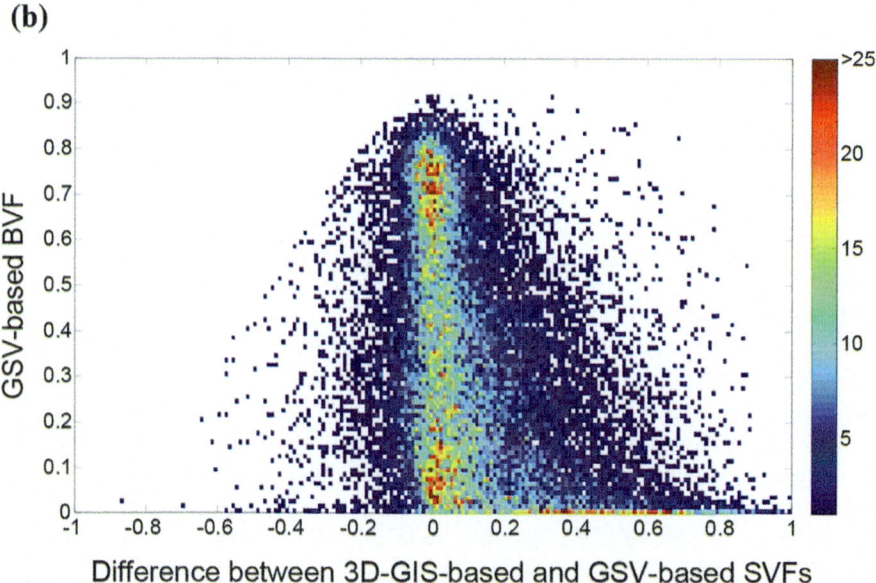

Fig. 4.7 (**a**) Bivariate histogram of GSV-based TVF and the error of SVF calculations from 3D-GIS, quantified using the difference between 3D-GIS-based and GSV-based SVFs. As indicated, when TVF > 0.1, the R^2 is 0.53 and the best fit linear slope is 1.17 (in dotted black line); (**b**) Bivariate histogram of GSV-based BVF and the error of SVF calculations from 3D-GIS. To derive the histogram, the data from both datasets are grouped into 0.01× 0.01 grids and the value for each grid is the total number of SVF samples that fall in the grid. The color shading indicates the number density and redder color indicates more data points at the grid. These two figures are used to quantify the impacts of TVF and BVF on the errors of SVF calculations from 3D-GIS

model simulations cannot parameterize street trees well due to their complexity, leading to underrepresentation of model simulations of realistic street environments. Therefore, in general, the 3D-GIS-based method produces larger sky openness and overestimate the SVF of street canyons in high-density urban areas in Hong Kong. This result differs from the study (Li et al., 2018) in the less dense Cardiff, UK, which indicates no significant correction between the difference in SVF estimates and street trees.

Figure 4.7b illustrates the bivariate histogram of GSV-based BVF, an indicator of building density, and the difference between the two SVF estimates. We can see that (1) most data differ between −0.1 and 0.1 and correspond to a wide range of BVFs from 0 to 0.9, and (2) when BVF is large, that is when building density is relatively large, the difference is centered at 0 with a small variation range, indicating good performances of model simulations in high-rise areas of Hong Kong. On the other hand, when BVF is small, the difference tends to be positive, indicating that the 3D-GIS-based SVF estimate is higher than the GSV-based estimate in lower density regions. Especially when BVF is close to 0, that is in the urban areas with very low building density, the differences can be very large, probably mainly due to the impact of trees, as shown in Fig. 4.7a. This figure shows that the lower the building density, the greater the difference. Combined with the effect of street trees on SVF values in Fig. 4.7a, our study shows that the larger amounts of street trees are associated with a higher uncertainty of modeled SVF. On the other hand, the higher the building density, the smaller the uncertainty.

4.4 Discussion

4.4.1 Large Uncertainty in Model-Based SVF Estimates from Street Trees

With the availability of 3D-GIS models for urban areas, the SVF can be continuously estimated at large spatial scales by simulating and calculating the projection of building blocks from any point on the ground. However, due to its complexity in shape and structure, street tree canopy information, a major feature of urban settings, is usually very difficult to parameterize and incorporate into models. As our study shows, the 3D-GIS-based method captures well the spatial pattern and variability of SVF in high-density urban areas of Hong Kong. However, it overestimates SVF by 0.11 on average. Moreover, our results show that as TVF increases by one unit, the resulting SVF error from the 3D-GIS-based method decreases by 1.17 unit, suggesting a significant correlation ($R^2 = 0.53$; $p < 0.01$) between street trees and the errors in model simulations. This result suggests that a lack of street trees is the dominant factor contributing to the large uncertainties in model simulations of urban street geometry. On the other hand, based on the linkage between street trees and the difference between 3D-GIS-based and GSV-based SVF estimates, the shade provision of street trees in the urban street canyon can be estimated using GSV images (Li et al., 2018). Street sensing images provide researchers with realistic obstructions along street canyons, i.e. buildings and trees, without only relying on simplifications or simulation models of the environment.

4.4.2 Temporal Variation of Street-Level View Factors

Hong Kong is located in a subtropical monsoon region with little effect of seasonality on the variation of the street tree canopy. An assumption on the seasonality is that specifically, the leaf cover of the street tree does not change during different seasons even though the acquisition time of GSV images differs (see Fig. 4.8). This is a reasonable assumption since Hong Kong is located in the subtropical monsoon region where the street trees can be maintained throughout the year (Jim, 1989). Moreover, Hong Kong is a highly developed high-density city where the built-up areas are limited and therefore very little change has taken place during recent years (Hong Kong Planning Department, 2015) that will significantly affect the street skylines.

However, for temperate climate regions, the seasonality of TVF will be a big issue, given that the street trees will be in an annual cycle of greening during growing seasons, and turning yellow and falling during the autumn and winter seasons. The change in color of tree leaves poses a challenge for VF study using conventional tree detection method based on the traditional spectral (RGB) information. The developed deep-learning method in this study, on the other hand, extract street features based on understandings of the local and global context information independent of spectral information, and therefore has an advantage over the traditional pixel-based spectral method. The developed method in this study can be used to address the problem of VF seasonality by first training the deep-learning module with tree image samples from different seasons, and then applying to GSV images grouped into different seasons.

The time information can be extracted from the GSV metafile as described in Sect. 3.3.1. This acquisition time of GSV image may enable us to understand the impact of seasons on the spatial pattern of VFs. Here we use Hong Kong as an example. For cities with seasonal trees, such as New York or Beijing, the time information is important to select and analyze the trees during the same season. Figure 4.8 shows the spatial distribution of the acquisition time of all the 22,729 GSV images collected in the high-density urban area of Hong Kong, including year in (a) and season in (b). 77.7% of the used GSV images in this study were updated and collected during the winter season from December 2016 to January 2017. Some GSV images have not been updated, which may be due to some uncontrolled factors (weather, road closures, etc.) such that Google cars may not be operating (Google Street View, 2017).

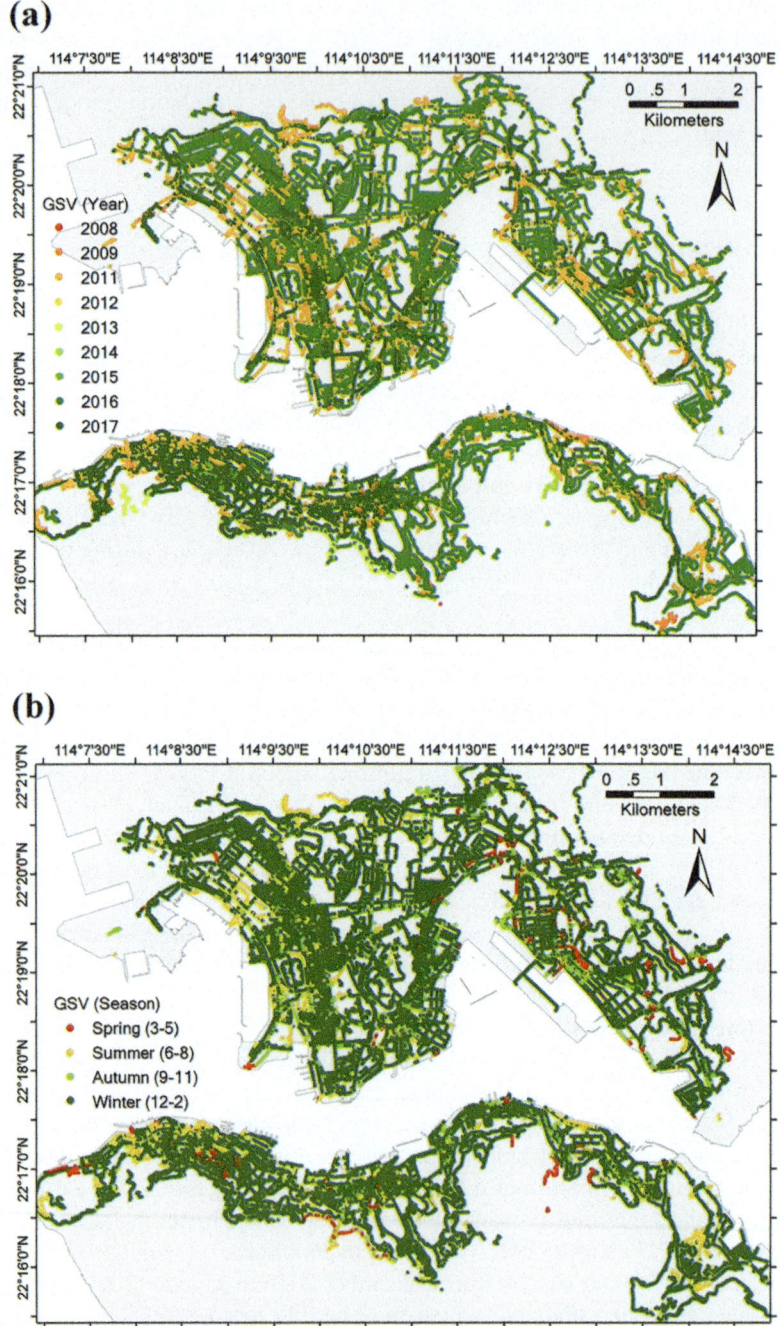

Fig. 4.8 Spatial distribution of the acquisition time, including year in (**a**) and season in (**b**), of the GSV images collected in the high-density urban area of Hong Kong in this study

4.4.3 Spatial Variation of Street-Level View Factors

The GSV-based method is applicable only in areas with GSV images for mapping streetscape variables, including SVF, BVF, and TVF. In areas without GSV data, such as the areas on Hong Kong Island, the 3D-GIS-based method or DSM-based method is still effective. The conclusions from studying the impact of street trees and building density on the uncertainty of the 3D-GIS-based method contribute to the future improvement of modeling the urban street environment. In this study, we use a 30-m interval for calculating VFs, assuming that this resolution would suffice to resolve the variation of VFs within a street. However, the GSV-based method is flexible in using any interval for mapping VFs of street canyons for study areas with different spatial scales.

GSV image may have the potential to investigate the symmetric and asymmetric characters of street canyons (Ali-Toudert & Mayer, 2007; Rodríguez-Algeciras et al., 2017) with the following extra inputs and assumption, including the exact location of observer, the width of the street which enables accurately calculating the observer position relative to both sides of the street, and assuming street trees don't block buildings on both sides. In this way, the height of buildings on both sides along the street can be calculated, and the asymmetric characters of street canyons may be determined.

4.5 Summary

This chapter focuses on (1) verifying the accuracy of the developed GSV-based method using reference data of hemispheric photography from field surveys; and (2) comparing the GSV-based and 3D-GIS-based VF estimates and investigating the impact factors for the discrepancies between them. As a result, maps of SVF, TVF, and BVF of street canyons in high-density urban areas of Hong Kong are generated. The following conclusions can be drawn:

- The spatial patterns of VF estimates are similar and consistent with the corresponding building height and density. The TVF is dominated by values less than 0.1 and the average value is 0.143, which is limited by the high building density and narrow street environment.
- Verification using reference data by hemispheric photography from field surveys shows that the GSV-based VF estimates have a satisfying agreement (with all R^2 values larger than 0.95) with the reference data. It suggests the effectiveness and high accuracy of the GSV-based method.
- A comparison between GSV-based and 3D-GIS-based SVFs show that the two SVF estimates are correlated (with R^2 of 0.40) and have a better agreement in high building density areas. However, the 3D-GIS-based method overestimates SVF by 0.11 on average.
- The differences between the two methods are significantly correlated with street trees ($R^2 = 0.53$). The more street trees, the larger the difference (by a factor of

1.17). This suggests that a lack of street trees in a 3D-GIS model of a street environment is the dominant factor, contributing to the large discrepancies between the two datasets. This study demonstrates an effective and accurate approach for mapping SVF in high-density areas of Hong Kong and suggests that street trees should be considered in model simulations of urban street environments.

In addition, sun view factor which is relevant to daytime shortwave irradiance can also be estimated by constructing geometries and orientations of street canyons using GSV images (Gong, 2019; Yang & Gong, 2025) in Chap. 5. The estimation involves the projection of the sun trajectory on the GSV fisheye image and then the calculation of the fraction of the length of the solar trajectory within the sky view range when sunlight can be seen. The quantification of street sun view factor by comparing the number of points of a sun path not blocked by obstacles with the total amount of points in the sun path in GSV images. It is critical for quantifying the sun-exposure of the solid surfaces which is needed to resolve the heterogeneity in urban areas for estimating thermal comfort more accurately.

References

Ali-Toudert, F., & Mayer, H. (2007). Effects of asymmetry, galleries, overhanging façades and vegetation on thermal comfort in urban street canyons. *Solar Energy, 81*(6), 742–754. https://doi.org/10.1016/j.solener.2006.10.007

Chen, L., Ng, E., An, X., Ren, C., Lee, M., Wang, U., & He, Z. (2012). Sky view factor analysis of street canyons and its implications for daytime intra-urban air temperature differentials in high-rise, high-density urban areas of Hong Kong: A GIS-based simulation approach. *International Journal of Climatology, 32*(1), 121–136. https://doi.org/10.1002/joc.2243

Gong, F.-Y. (2019). *Mapping street canyon morphology and solar radiation in high-density urban environments using street sensing approach*. The Chinese University of Hong Kong.

Gong, F.-Y., Zeng, Z.-C., Zhang, F., Li, X., Ng, E., & Norford, L. K. (2018). Mapping sky, tree, and building view factors of street canyons in a high-density urban environment. *Building and Environment, 134*, 155–167. https://doi.org/10.1016/j.buildenv.2018.02.042

Google Street View. (2017). *Google street view—Where we've been & where we're headed next*. Retrieved January 18, 2018, from https://www.google.com/streetview/understand/

Hong Kong Planning Department. (2015). *Planning department - Hong Kong planning standards and guidelines - Contents*. Retrieved November 2, 2018, from https://www.pland.gov.hk/pland_en/tech_doc/hkpsg/full/index.htm

Jim, C. Y. (1989). Tree-canopy characteristics and urban development in Hong Kong. *Geographical Review, 79*(2), 210–225. https://doi.org/10.2307/215527

Johnson, G. T., & Watson, I. D. (1984). The determination of view-factors in urban canyons. *Journal of Climate and Applied Meteorology, 23*(2), 329–335. https://doi.org/10.1175/1520-0450(1984)023<0329:TDOVFI>2.0.CO;2

Li, X., Ratti, C., & Seiferling, I. (2018). Quantifying the shade provision of street trees in urban landscape: A case study in Boston, USA, using Google street view. *Landscape and Urban Planning, 169*(Supplement C), 81–91. https://doi.org/10.1016/j.landurbplan.2017.08.011

MIT Senseable City Lab. (2016). *Treepedia: MIT senseable city lab*. Retrieved November 24, 2017, from http://senseable.mit.edu/treepedia

References

Oke, T. R. (1987). *Boundary layer climates*. Routledge.

Rodríguez-Algeciras, J., Tablada, A., & Matzarakis, A. (2017). Effect of asymmetrical street canyons on pedestrian thermal comfort in warm-humid climate of Cuba. *Theoretical and Applied Climatology, 133*, 663–679. https://doi.org/10.1007/s00704-017-2204-8

Yang, Z., & Gong, F. Y. (2025). Utilizing street view images to estimate solar energy potential for photovoltaic powered buses. *Applied Geography, 177*, 103567. https://doi.org/10.1016/j.apgeog.2025.103567

Chapter 5
Spatiotemporal Patterns of Street Canyon Solar Radiation

Contents

5.1	Overview..	70
5.2	Methods Verification...	71
	5.2.1 Accuracy Assessment of GSV-Based Solar Radiation Method.....................	71
	5.2.2 Verification of GSV-Based Street-Level Solar Radiation Estimates...............	73
	5.2.3 Verification of GSV-Based Free-Horizon Solar Radiation Estimates..............	74
5.3	Results..	77
	5.3.1 Spatiotemporal Pattern of Clear-Sky Street-Level Solar Irradiation...............	77
	5.3.2 Spatiotemporal Pattern of all-Sky Street-Level Solar Irradiation....................	78
	5.3.3 Contributions from Direct and Diffuse Components.....................................	81
	5.3.4 Effect of Street Canyon Geometry on Solar Irradiation.................................	82
5.4	Discussion...	85
	5.4.1 Spatial Inhomogeneity of Solar Radiation...	85
	5.4.2 Reflected Radiation in a Street Canyon and its Impact.................................	86
	5.4.3 Transmissivity of Solar Radiation Through Tree Crowns..............................	89
5.5	Summary..	89
References...		90

Abstract This chapter presents a methodology to assess street-level solar irradiation in high-density urban environments using Google Street View (GSV) images, validated against field measurements and Hong Kong Observatory (HKO) data. Three key factors—solar geometry, atmospheric conditions (e.g., cloudiness), and street canyon morphology (sky view factor, orientation)—are integrated to quantify spatiotemporal patterns. Verification shows strong agreement between GSV-derived estimates and observational data (correlation coefficients: 0.86–0.98), confirming the model's accuracy in capturing diurnal and seasonal cycles. Results demonstrate significant seasonal variation: summer irradiation exceeds winter by a factor of 3–5, driven by solar zenith angle changes and cloud cover. Spatially, low-rise areas (sky view factor ≥0.7) receive 3× more irradiation than high-rise zones (sky view factor ≤0.3), with West-East-oriented streets experiencing prolonged solar exposure in summer. Diffuse radiation dominates in winter, while direct radiation contributes 70% of summer totals. The study highlights the role of urban geometry, showing that high aspect ratios (H/W ≥ 2) reduce solar access by 80% in winter. Spatial inhomogeneity analysis reveals <10% variation in regional solar radiation, justifying the use of centralized HKO data. Limitations include unmodeled reflections and tree transmissivity, warranting future refinements. These findings provide critical insights for urban energy management and microclimate mitigation.

Keywords High-density urban environments · Solar irradiance estimation · Google street view (GSV) · Street canyon geometry · Spatiotemporal variability

5.1 Overview

There are three key factors influencing the street-level solar energy in a high-density environment: (1) Solar geometries including solar zenith angle (SZA) and solar azimuth angle (SAA). In Hong Kong (22°17′7.87″N, 114°9′27.68″E), the SZA at local noon time changes from 45.5° in the winter and 3.5° in the summer. The solar geometries determine the associated solar irradiance on a horizon surface due to the seasonal variation; (2) Atmosphere conditions including cloudiness, local atmospheric pressure, and Linke turbidity factor. The obvious influence of street-level solar radiation comes from the cloud, because of its effects on obstructing and reflecting the solar radiation back to other space, as well as creating the diffuse radiations into street spaces. Relative to the strong influence of clouds, the local atmospheric pressure and Linke turbidity factors influence the solar energy associated with their magnitude of absorption which has a seasonal effect; (3) Street canyons geometry, including street orientation and sky view factor (SVF). The SVF is the fraction of sky in the upper hemisphere of a street canyon and determines the total incoming solar radiation that reaches the top of a street. The orientation of street canyons affects the timing of exposure to direct sunlight, particularly in a high-density environment with high H–W ratio.

This chapter describes an approach for calculating solar irradiance of street canyons using Google Street View (GSV) images and investigates its spatiotemporal patterns in a high-density urban environment. In this method (Gong, 2019; Gong et al., 2018; Gong et al., 2019), GSV images provide a unique way to characterize the street morphology from which the diurnal solar path and solar radiation exposure can be estimated in a street canyon. Verifications of our developed method using free-horizon HKO observations and street-level field measurements show that both the calculated clear-sky and all-sky solar irradiance of street canyons well capture the diurnal and seasonal cycles.

5.2 Methods Verification

5.2.1 Accuracy Assessment of GSV-Based Solar Radiation Method

To evaluate the accuracy of the developed GSV-based street-level solar irradiance estimation method, two ways of verifications are implemented: (1) street canyon verification using field measurements by LI-200R Pyranometer in a high-density environment; (2) free-horizon verification using HKO measurements of direct and diffuse radiation at KP Meteorological Station in Kowloon (Hong Kong Observatory, 2003c) and KSC Solar Station in Sai Kung (Hong Kong Observatory, 2012).

Firstly, field measurements are conducted in May 2018 to obtain the solar radiation of street canyon in a high-density high-rise street environment. The measurement is located on the campus of The Chinese University of Hong Kong, as shown in Fig. 3.1b. For the purpose of model validation, field measurements were conducted between 05:00 and 20:00 Hong Kong local time on two clear days, May 22 and May 23, 2018. A LI-200R Pyranometer measured global solar irradiance, i.e., the combination of direct and diffuse solar radiation in the 400–1100-nm range (LI-COR Biosciences, 2015). Table 5.1 summarizes the atmospheric conditions during field measurement days. The results are shown in Sect. 5.2.2.

Table 5.1 describes the associated atmospheric conditions and site information for field measurements. The temporal sampling interval and measurement time are also indicated. Figure 5.1 shows the measured free-horizon global, direct and diffuse solar irradiance. According to Li and Lam (2002), the clear day criteria are given by: (a) The direct normal irradiance is greater than 200 W/m^2; (b) The ratio of the diffuse component to global irradiance should be less than 1/3. From the results of free-horizon solar irradiance measurements in Fig. 5.1, we can see May 22–23, 2018 meet the criteria for clear days.

Secondly, free-horizon verification using HKO measurements of direct and diffuse radiation at KP Meteorological Station in Kowloon (Hong Kong Observatory, 2003c) and KSC Solar Station in Sai Kung (Hong Kong Observatory, 2012). Global

Table 5.1 Summary of time information and site information of street canyon field measurement, and associated atmospheric conditions used in the model

Time information			Atmospheric conditions		
Date	Time (h)	Interval	Highest free-horizon solar irradiance (W/m^2)	Minimum solar zenith angle (°)	Mean sea-level pressure (hpa)
22-May-2018	05:00–22:00	15 s	997.3	4.2	1010.5
23-May-2018	05:00–22:00	15 s	963.6		1009.6
Site information			View factor (VF)		
Longitude	Latitude	Altitude	SVF	TVF	BVF
114°12′36.1″E	22°25′06.1″N	50.0 m	0.47	0.13	0.40

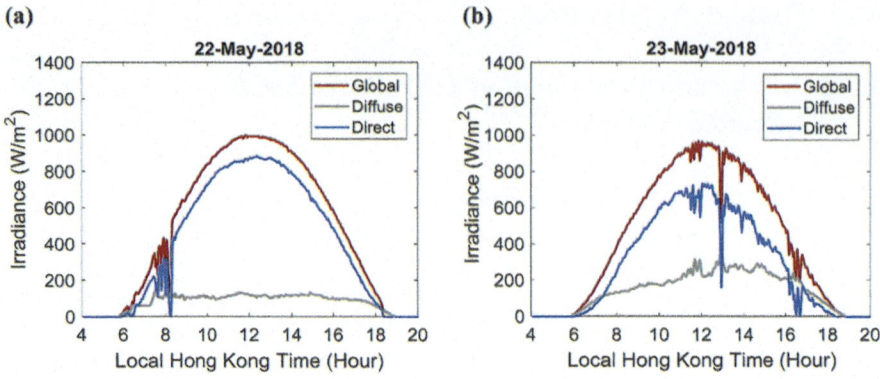

Fig. 5.1 The free-horizon solar irradiance measurements by HKO for 22 May 2018 in (**a**) and 23 May 2018 in (**b**)

Fig. 5.2 King's Park and Kau Sai Chau meteorological stations. (**a**) Solar radiation measuring equipment at the King's Park (KP) HKO Station (Hong Kong Observatory, 2003c). (**b**) Direct and diffuse solar radiation instruments (left) mounted on a sun tracker and the global solar radiation sensor (right) in Kau Sai Chau solar station

radiation was first measured at the Royal Observatory Kowloon in 1957. Form 1960 it has been measured at King's Park (KP), and located at 22°19′N, 114°10′E, 65 m MSL. Currently, direct and diffuse solar radiation measured by the Hong Kong Observatory at King's Park and Kau Sai Chau meteorological and solar stations (indicated by stars in Fig. 3.1b). A pyranometer is used to measure diffuse solar radiation, as shown in Fig. 5.2, by shading direct radiation from the sun; a pyrheliometer mounted on a sun tracker is used to measure direct solar radiation. The sun tracker is used to ensure that the pyrheliometer is pointing directly at the sun (Hong Kong Observatory, 2012). The direct solar radiation is converted to that over a horizontal surface by multiplying the cosine of solar zenith angle. A real-time measurement can be seen in (Hong Kong Observatory, 2003b). Kau Sai Chau, which is located in a more rural setting, has a slightly higher amount of global and direct solar radiation in most months than King's Park in a more urban environment. On

5.2 Methods Verification

the other hand, King's Park has a slightly higher amount of diffuse solar radiation in early months and a slightly lower amount in late months.

In this study, the solar radiation measurements in KP is firstly used as top-of-roof solar energy to calculate the all-sky street canyon solar energy given it is effectively located in the center of our study area (Sect. 5.2.2); secondly, HKO measurements are used to verify the calculated clear-sky solar radiation in free-horizon (Sect. 5.2.3); finally, the difference between KP and KSC is used to characterize the spatial heterogeneity of incoming solar energy over this region (Sect. 5.4.1).

5.2.2 Verification of GSV-Based Street-Level Solar Radiation Estimates

Calculation of all-sky street-level solar irradiance developed in this study based on GSV images and HKO measurements is verified using field measurements in a high-density urban street canyon. Figure 5.3 presents the measured and simulated global radiation for two consecutive days, May 22–23, 2018. Figure 5.3a shows the fisheye image of the street canyon with the corresponding solar path over a day. The panorama classification is shown in Fig. 5.3b for the sky, tree, and building, of which the SVF, TVF, and BVF are 0.47, 0.13, and 0.40, respectively. The sun path is obstructed by trees in the eastern section of the street canyon in the morning before ~8:00 h and by the buildings in the western section of the street canyon in the afternoon after 13:00 h. Between 8:00 h and 13:00 h, when the sky is open, we can see the calculated global irradiance fits very well with the street canyon field measures and the free-horizon HKO data, as shown in Fig. 5.3c, d. The small spikes in field measurements during 10:30–11:00 h is due to the specular reflectance from the building windows, which is hard to capture in the calculation. The sharp drawdown at around 13:00 h is very likely due to clouds. During 6:30–8:30 h, the field measurement data is slightly higher than the GSV-based estimated radiation. This is probably because the sunlight is not totally blocked by the trees and there is still some transmissivity of solar radiation through the tree crowns. When the sun path is obstructed, the solar radiation decreases dramatically relative to free-horizon radiation and is reduced to be only from diffuse radiation.

We can see the calculations well capture the temporal pattern of the diffuse radiation. The differences between calculated and measured global irradiance range from -20 to $+20$ W/m^2 during sunshine hours. These differences can be partly explained by the effect of reflected radiation within the street canyon. The differences increase to more than 20 W/m^2 in the afternoon when there are clouds. Figure 5.3e, f shows that the correlation coefficients of estimated and measured global irradiance are 0.86 and 0.98, respectively. The outlier points off the diagonal line are due to clouds or specular reflectance. From these results, we can conclude that the general pattern of the solar irradiance in a street canyon can be predicted by the developed GSV-based model.

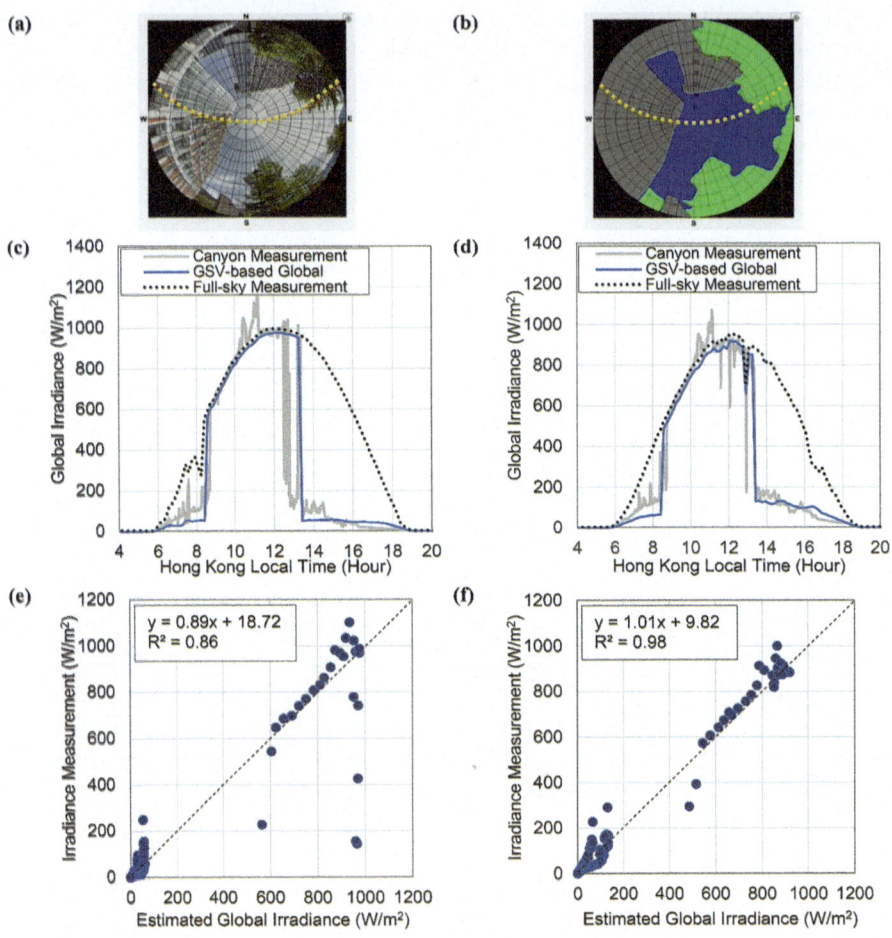

Fig. 5.3 Comparison of measured and GSV-based estimated global irradiance in the street canyon measurement site located in the campus of The Chinese University of Hong Kong. (**a**) Fisheye image of the street canyon; (**b**) Sky, buildings, and trees features of the street canyon; Comparison of measured and GSV-based estimated global irradiance between 05:00 h and 20:00 h for 22 May 2018 in (**c**) and 23 May 2018 in (**d**); Scatter plots between GSV-based estimated and field measured global irradiance from 5:00 h to 20:00 h for 22 May 2018 in (**e**) and 23 May 2018 in (**f**)

5.2.3 Verification of GSV-Based Free-Horizon Solar Radiation Estimates

Further verification against HKO measurements under free-horizon view is described in this section. Multi-year HKO solar irradiance measurements in free horizon provide an essential way to verify our proposed method for estimating street-level all-sky irradiance under all cloud conditions. The hourly mean of free-horizon solar irradiance is calculated for the King's Park site with associated atmospheric conditions, including sea-level pressure (see Fig. 5.4a), Linke turbidity factor (see Fig. 3.4), and cloudiness from HKO (see Fig. 5.4b). The results are then

5.2 Methods Verification 75

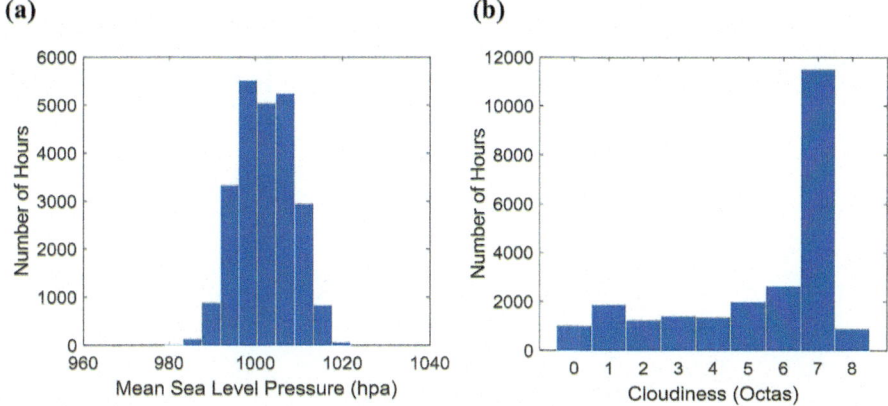

Fig. 5.4 The histograms of hourly averaged sea-level pressure in (**a**) and cloudiness in (**b**) between 8:00 h and 18:00 h from HKO measurements from 2009 to 2014

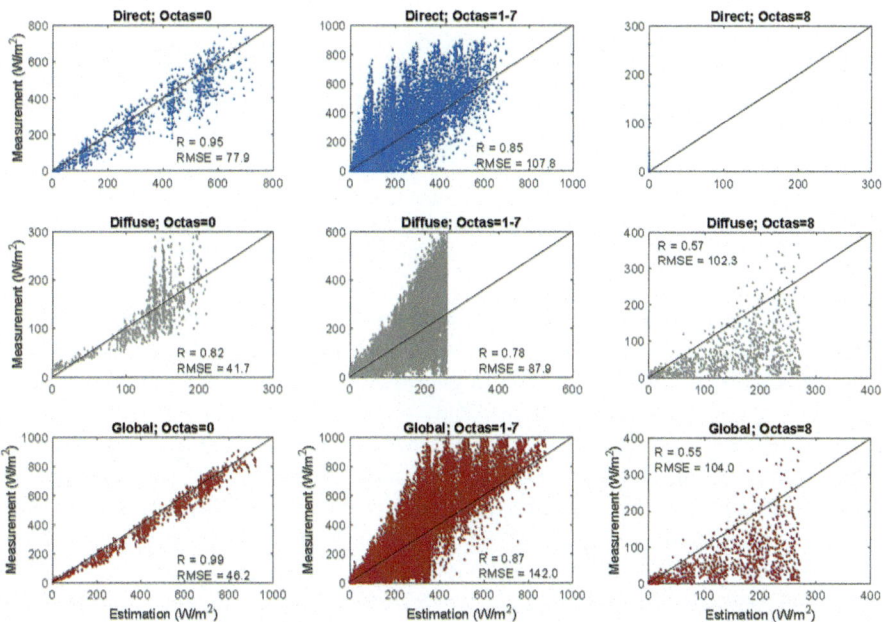

Fig. 5.5 Scatter plots between the calculated free-horizon hourly solar irradiance and HKO measurements at the site of King's Park. The plots are grouped by hourly solar irradiation, including direct and diffuse components, under different cloud coverages in terms of octas as measured for 6 years by HKO from 2019 to 2014

compared with HKO solar irradiance measurements by investigating their difference. Given the large uncertainty from the cloud effect, the differences in direct, diffuse and global solar irradiance are plotted for three sets of different cloud coverages, as shown in Figs. 5.5 and 5.6. In total, 2190 days of data from 2009 to 2014 are used. As shown in Fig. 5.5, the case for clear days (octas = 0) has the best

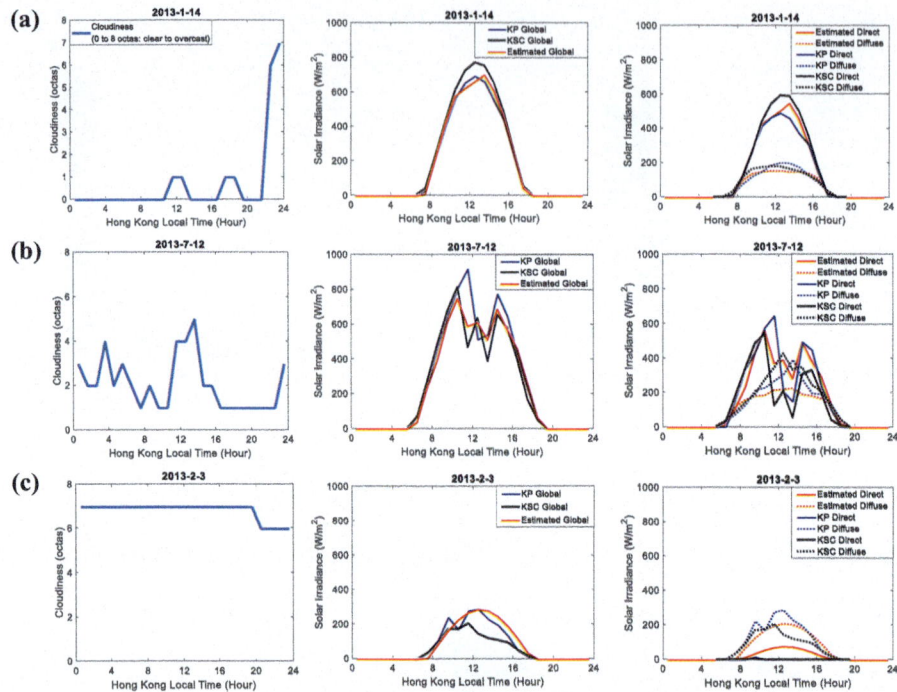

Fig. 5.6 Comparison of estimated all-sky irradiance and HKO measurement of hourly global, direct, and diffuse solar irradiance for three different cloudiness example days: (**a**) A clear day (octas = 0–1); (**b**) A semi-cloudy day (octas = 2–6); (**c**) A cloudy day (octas = 7–8)

accuracy with the highest correlation coefficient (0.99) and the smallest RMSE (46.2 W/m^2) for global irradiance. This result justifies the method described in Sect. 3.4.3 to calculate the clear-sky solar radiation under clear-sky assumptions. In general, for the semi-cloudy condition (octas = 1–7), the calculated solar irradiance tends to underestimate the HKO measurements, while for the overcast condition (octas = 8), the calculated solar irradiance tends to be overestimated. Results from further statistical analysis demonstrate that: (1) the overall correlation coefficient for global solar irradiance under all-sky conditions (for all octas from 0 to 8) is 0.87 and the RMSE is 138.2 W/m^2; (2) about 47.6% of the solar irradiance calculations under all cloud conditions (octas from 0 to 8) have errors smaller than 50 W/m^2. The accuracy of the calculation method under partially cloudy conditions remains to be improved, given that current parameterized models cannot account for cloud movement across the sky that may create complex patterns of reduced and enhanced radiation values.

Three examples with clear, semi-cloudy, and overcast days are shown in Fig. 5.6 to investigate the diurnal variability of calculated solar irradiance under different cloudiness conditions compared with HKO measurements. The global, direct, and diffuse solar irradiance are illustrated. We can see in general the calculated solar irradiance well captures the diurnal variabilities of the measured HKO solar irradiance on the clear day. However, there are slight differences in the cases of semi-cloudy day and overcast day. Differences between HKO measurements at KP and KSC sites are due to the inhomogeneity of solar radiation probably related to different atmospheric conditions among this region. An investigation of this inhomogeneity is presented in Sect. 5.4.1.

5.3 Results

In this section, street-level solar radiation is calculated using the GSV-based method under two conditions: (1) clear-sky solar irradiation under ideal clear-sky assumptions to investigate the effects of street morphologies and geometries; and (2) all-sky solar irradiation based on HKO measurements to investigate the impact of clouds.

5.3.1 Spatiotemporal Pattern of Clear-Sky Street-Level Solar Irradiation

Clear-sky street-level solar irradiation is the solar radiation in street canyons under clear day (no cloud) assumptions. Its spatial and temporal patterns are therefore dominantly affected by solar geometries, street canyon geometries (street orientation and aspect ratio), and morphologies (sky opening and obstructions by buildings and trees). Figure 5.7 shows the monthly mean of daily clear-sky solar irradiation averaged over 6 years from 2009 to 2014 in the high-density urban areas of Hong Kong. A strong seasonal variation with high values in the summer and low values in the winter can be observed. This temporal pattern mainly follows the seasonal variation of incident solar zenith angle. From Fig. 5.7, two distinct spatial features can be seen: (1) the spatial variability is closely related to building densities as shown in Fig. 3.1c, in which much lower solar radiation is received in streets surrounding by high-density buildings. A similar pattern of spatial variation can also be seen in GSV-based SVF estimates as shown in Fig. 4.5a; (2) streets with West-East orientation receive higher solar radiation than surrounding regions in summer when SZA is small and the street is exposed to solar radiation for the whole day. In winter, the SZA is large and therefore most of the direct solar irradiation is obstructed by buildings. Further analysis of street-level irradiation with different street canyon morphologies is presented in the following Sect. 5.3.4.

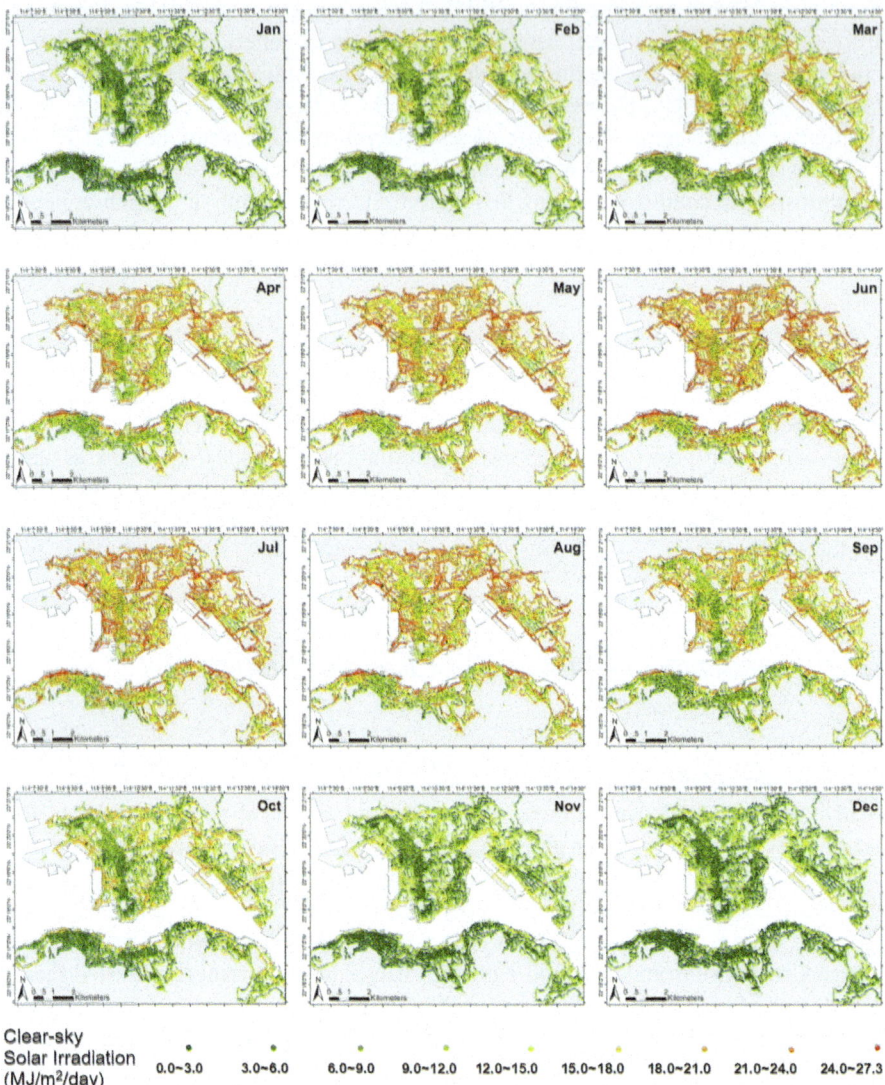

Fig. 5.7 Monthly mean of daily clear-sky solar irradiation (MJ/m²/day) in street canyons averaged over 6 years from 2009 to 2014 in the high-density urban areas of Hong Kong. Calculation of the clear-sky solar irradiation is introduced in Sect. 3.4.3

5.3.2 Spatiotemporal Pattern of all-Sky Street-Level Solar Irradiation

Figure 5.8 shows the corresponding monthly mean all-sky solar irradiation of street canyons calculated using GSV images and realistic HKO measurements as described in Sect. 3.4.3. The spatial pattern of the all-sky solar irradiation shown in Fig. 5.8 is similar to those in the clear-sky solar irradiation as shown in Fig. 5.7. Similarly, the

5.3 Results

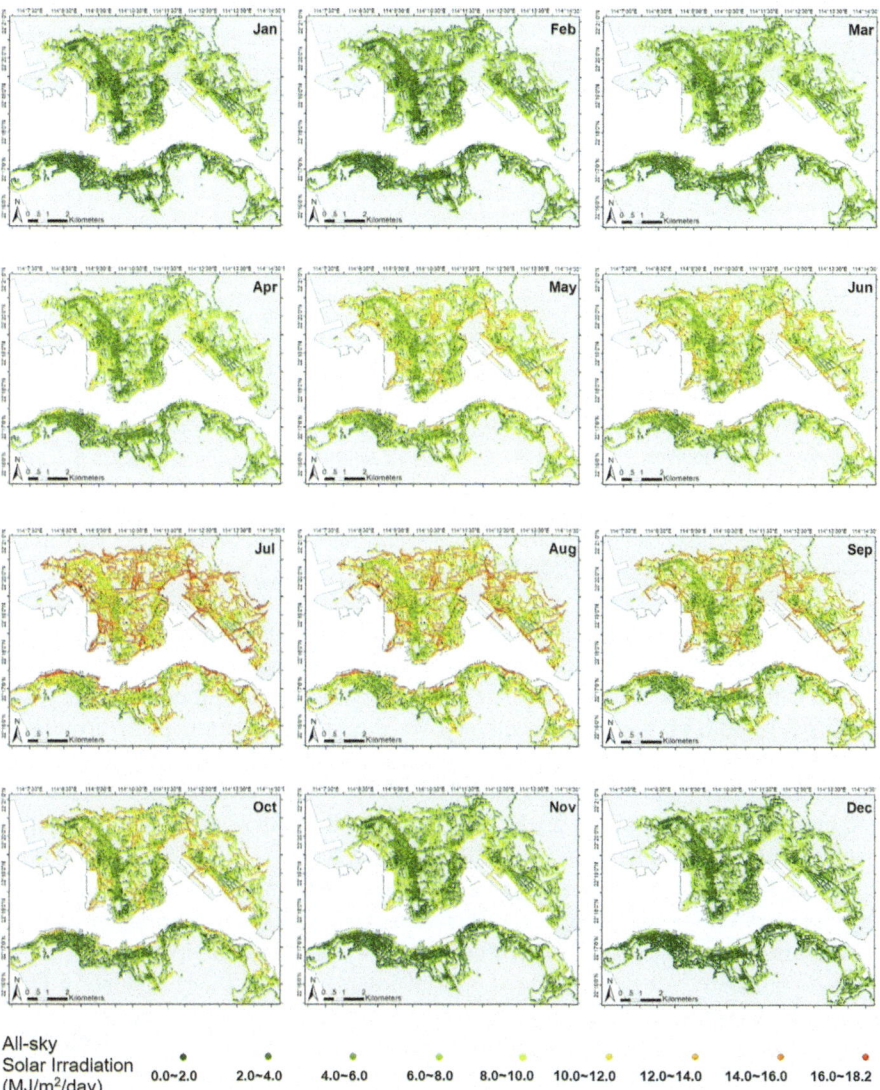

Fig. 5.8 Same as Fig. 5.7 but for all-sky street-level solar irradiation. These monthly means of daily all-sky solar irradiation (MJ/m^2/day) in street canyons are averaged over 6 years from 2009 to 2014 in the high-density urban areas of Hong Kong. The detail of the calculation is introduced in Sect. 3.4.3

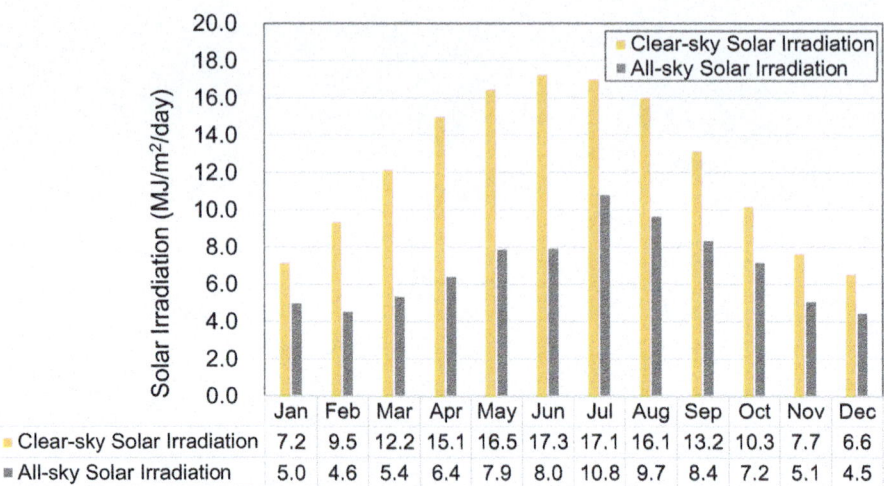

Fig. 5.9 Comparative analysis of monthly mean of the calculated daily clear-sky and all-sky solar irradiation (MJ/m²/day) based on 25,654 street canyon samples using Google Street View images in the high-density urban area of Hong Kong

street geometries and morphologies have dominant imprints on the spatial pattern of the all-sky solar irradiation. To quantitatively compare the street-level solar irradiation in extreme high-rise and low-rise street canyons, we calculate the means with SVF ≥ 0.7 and SVF ≤ 0.3,[1] respectively. We found that in summer, the irradiation in an extreme low-rise region (16.1 MJ/m²/day) is on average about three times higher than that in extreme high-rise region (5.6 MJ/m²/day). In winter, the low-rise and high-rise irradiations are 9.1 MJ/m²/day and 1.8 MJ/m²/day on average, respectively, differing by a factor of 5.

As shown in Fig. 5.9, when realistic cloud effect is considered in the calculation, the solar irradiation value drops significantly, especially from February to June the values are less than half of the corresponding clear-sky solar irradiation. This is because clouds cover 80% of the sky on average from February to June in Hong Kong, according to the long-term cloud data from Hong Kong Observatory (2010). During the summer season, the dominant weather condition is cloudy, especially in the early afternoon (Giridharan et al., 2008). Cloudy conditions, with cloud amount reaching 6 octas[2] or above, occurred during about 70% of the summer, mostly between 12:00 h and 15:00 h in the afternoon (Hong Kong Observatory, 2018).

---

[1] The high-rise, mid-rise, and low-rise regions in Hong Kong urban areas can be identified in our study by SVF ≤ 0.3, 0.4 ≤ SVF ≤ 0.6, and SVF ≥ 0.7, respectively, according to local climate zone classification (Stewart & Oke, 2012).

[2] The reported or forecast cloud amount is coded in accordance with the following five categories: SKC (sky clear, representing 0 oktas); FEW (few, 1–2 octas); SCT (scattered, 3–4 octas); BKN (broken, 5–7 octas); and OVC (overcast, 8 octas). (Source: Hong Kong Observatory, Technical Note No. 105. Verification of Weather Forecasts for the Aerodrome of the Hong Kong International Airport. 2003. http://www.hko.gov.hk/publica/tn/tn105.pdf).

From Fig. 5.9, we can see that the lowest and highest monthly mean solar irradiation are 4.5 MJ/m² in December and 10.8 MJ/m² in July. As shown in Fig. 5.9, the lowest monthly averaged value is ~6.6 MJ/m²/day in December and the highest value is ~17.2 MJ/m²/day in July. Higher daily solar irradiation of street canyons (>20.0 MJ/m²/day) occurred during the summer season is mainly distributed in open space areas (SVFs are around 0.7–1.0) or street canyons with West-East orientation (SVFs are around 0.3–0.6). Lower solar irradiation of street canyons (<10.0 MJ/m²) during most of the winter season is mainly distributed in high building density areas (BVFs are around 0.7–0.8) or high tree coverage areas (TVFs are around 0.6–0.8).

5.3.3 Contributions from Direct and Diffuse Components

For each specific street canyon, the horizon obstructions (e.g., buildings and trees) will exert independent effects on the direct and diffuse radiation. When the sun is masked by an obstacle, such as buildings or trees, the direct radiation in Eq. (3.7) and the anisotropic diffuse radiation in Eq. (3.8) will be totally blocked. The remaining portion of solar diffuse radiation, i.e., the isotropic diffuse radiation, enters the street through the sky opening. For cloudless skies, direct irradiance in street canyons is determined by whether the solar disk is obstructed or not, while diffuse radiation in street canyons is proportional to the amount of sky opening, i.e., SVF. For this reason, a detailed study of the spatial patterns of direct and diffuse irradiation provides further clues on how street morphologies affect the two components of street-level solar irradiation. To calculate their contributions to the total irradiation, it is first necessary to split the global radiation into the direct and diffuse components, as shown in Eq. (3.6).

Figure 5.10 illustrates the direct and diffuse irradiation separately for winter (January) and summer (July) averaged over 6 years from 2009 to 2014. For the direct irradiation, shown in Fig. 5.10a, c for winter and summer, respectively, the spatial patterns are similar but the value in summer is about two times higher than that in winter, especial in street canyons with East–West orientation. This is because the SZA is summer is much smaller than that in winter. Moreover, there is less obstruction by buildings or trees in summer when SZA is smaller. For the diffuse irradiation, there are large differences in spatial patterns in the two seasons. In summer days, as expected, the areas have higher diffuse irradiation than that in winter days, as shown in Fig. 5.10b, d, because the anisotropic diffuse component in summer, when the sun path is less obstructed, is higher compared to that in winter. For the direct and diffuse irradiations in winter as shown in Fig. 5.10a, b, the value in the high-density regions is very low. The direct component is higher than the diffuse component in the open areas with high SVF. However, the street orientation has a relatively small effect on received solar radiation in winter because of a relatively short exposure time of direct solar irradiation due to sunlight obstructions.

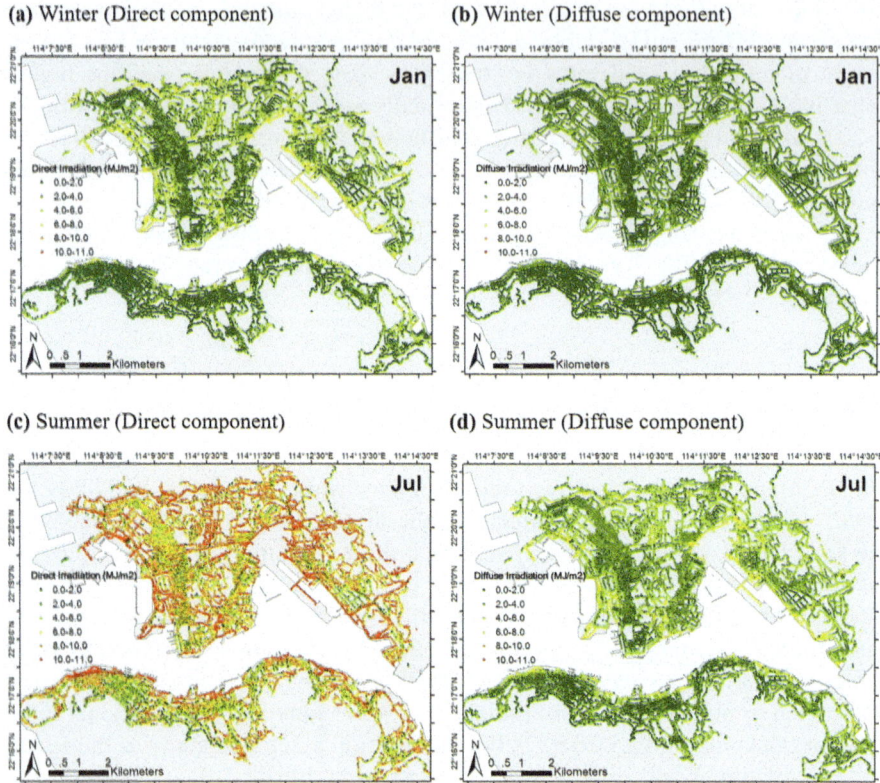

Fig. 5.10 (**a**) Daily direct irradiation (MJ/m^2/day) of street canyons in high-density urban areas of Hong Kong in January averaged for 6 years (2009–2014); (**b**) Daily diffuse irradiation (MJ/m^2/day) of street canyons in January averaged for 6 years (2009–2014); (**c**) the same as (**a**) but for July, an example of summer; (**d**) The same as (**b**) but for July. These are all-sky street-level solar irradiation estimates based on GSV images and HKO measurements as described in Sect. 3.4.3

5.3.4 Effect of Street Canyon Geometry on Solar Irradiation

An investigation of the variation of street-level solar irradiation between different seasons is directly linked to the street geometry and morphologies and therefore helps identify regions with solar radiation over-exposure in summer but under-exposure in winter. Figure 5.11 shows the coefficient of variation, defined as the ratio of standard deviation to mean, of daily solar irradiation calculated using 2190 days of all-sky solar irradiation data over the 6 years from 2009 to 2014. We can see that high variability is mainly located in high-density street canyons,

5.3 Results

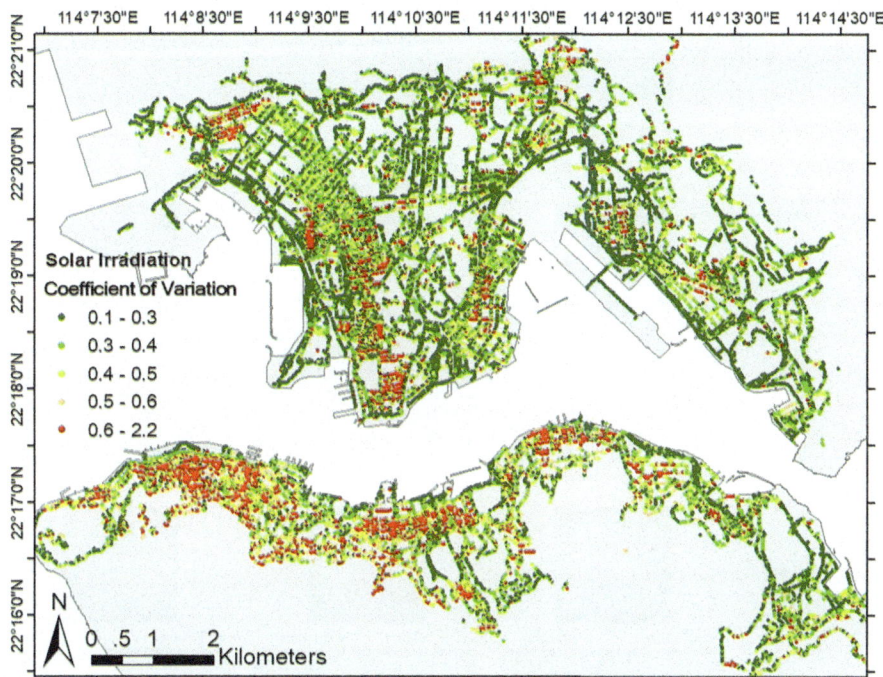

Fig. 5.11 Coefficient of variation, defined as the ratio of standard deviation to mean, of daily solar irradiation calculated using 2190 days of all-sky solar irradiation data over the 6 years from 2009 to 2014

especially in streets with West–East orientation, while low variability can be seen at low-density build-up areas and streets with non-W-E street orientation. This agrees with the month to month variability as shown in Fig. 5.9.

To further examine the dependence of solar irradiation on urban street geometries and morphologies, which are characterized by street orientation and H/W ratio, we select six different examples of street canyons, including three different types of street geometries with H/W ratios of 1/2, 1, and 2, respectively, and two different types of street orientations of North–South and West–East, as shown in Fig. 5.12. The SVF of these street canyons changes from 0.2 to 0.8, and the daily clear-sky solar irradiation changes from about 1.0 MJ/m^2/day in winter to more than 16 MJ/m^2/day in summer. Figure 5.13 shows the monthly mean of the daily all-sky global, direct, and diffuse irradiation corresponding to the six types of street canyons.

As expected, solar irradiation increases as SVF increases from high-rise (H/W $\geq$ 2:1) to low-rise (H/W $\leq$ 1:2). The important role of street orientation in the incident solar irradiation can be observed in street canyons of West–East and South–North

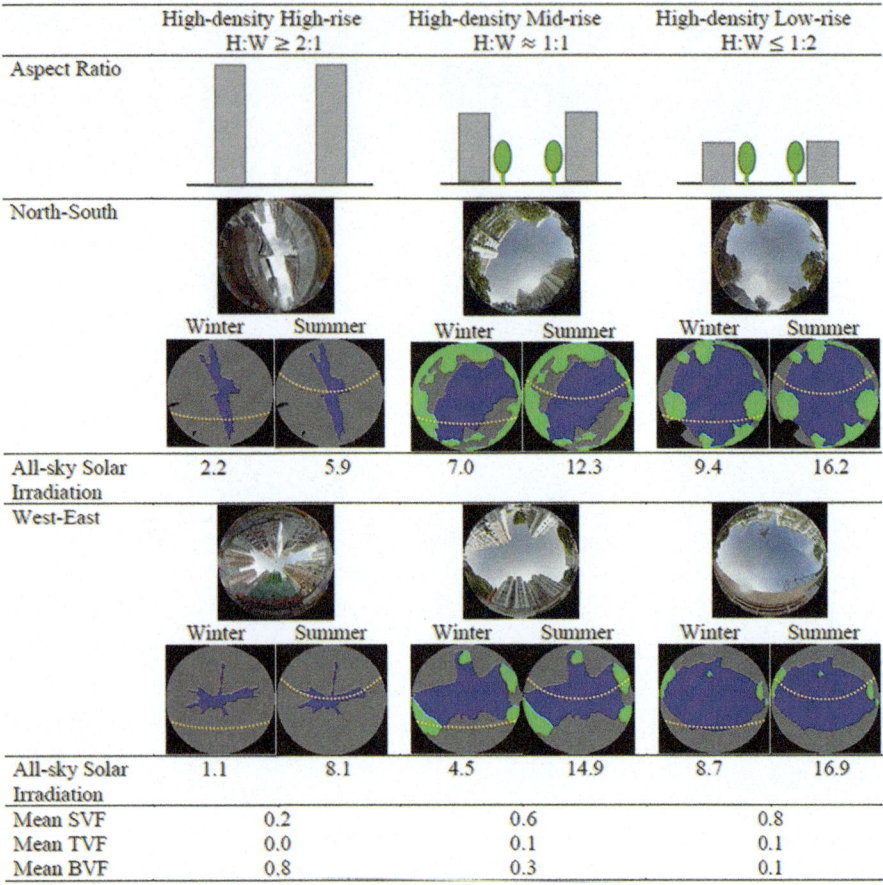

Fig. 5.12 Six different types of street canyons, including three different types of street geometries with H/W ratios of 1/2, 1, and 2, respectively, and two different types of street orientations of North–South and West–East. The daily all-sky solar irradiation (MJ/m²/day) for summer and winter are also indicated, respectively, as well as sky view factors (SVF), tree view factors (TVF), and building view factor (BVF)

orientations with the same aspect ratio. In general, street canyons with West–East orientation receive higher solar radiation in the summer season from March to September but lower in the winter season.

This difference results from the change of solar incident zenith angle between summer and winter. When SZA is small in summer, the West–East orientation streets are exposed to solar radiation for a much longer time from morning to afternoon compared to North–South orientation street while in winter when SZA is high, most solar radiation is obstructed by buildings and trees for a West-East orientation street. However, the North–South orientation street can still get exposure to solar

5.4 Discussion

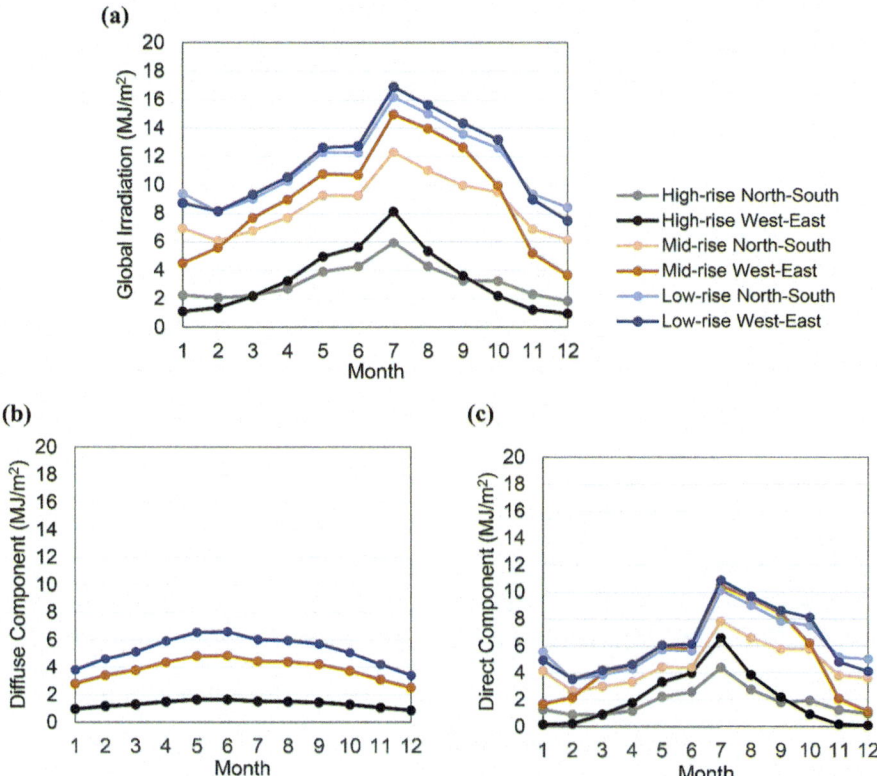

Fig. 5.13 Monthly mean of daily all-sky solar irradiation (MJ/m^2): (**a**) global irradiation, (**b**) diffuse irradiation and (**c**) direct irradiation for the six types of street canyons in high-density urban areas of Hong Kong averaged for 6 years from 2009 to 2014

radiation at noon. As shown in Fig. 5.13b, c, this difference is obvious in total solar irradiance and direct component, but not in the diffuse component, which is nearly similar for different street orientations.

5.4 Discussion

5.4.1 Spatial Inhomogeneity of Solar Radiation

In this study, the solar irradiance measurements in KP, which is located in the middle of the study area as shown in Fig. 3.1b, are assumed to represent the whole study area and are used in Eqs. (3.7) and (3.8) to calculate the all-sky street-level solar irradiance. Here we compare the measurements from KP and KSC sites to justify this assumption. KSC is located to the North–East of the study area (see Fig. 3.1b) and the distance between KSC and KP is about two times the width/length of the study area. The difference between the two characterizes the spatial homogeneity of incident solar radiation over this region. As shown in Fig. 5.14a, in general, the global radiation at KP agrees closely with that in KSC. KSC, a more rural setting, has a slightly higher amount of direct solar irradiation in most months than KP in a more urban environment, probably due to the stronger effect of aerosol extinction in KP. The largest difference in the summer months is about 10% in global radiation. The diffuse radiation is almost the same between KP and KSC, indicating the cloud diffusion effect is not causing bias in spatial distribution over this region. These results imply that the spatial difference in incident global solar radiation is small over the whole Hong Kong territory. A similar conclusion can be reached from Fig. 5.14b, more than 90% of the days have a difference less than 2.0 $MJ/m^2/day$ for both direct and diffuse components at street level. Within the uniformly high-density urban study area, the difference should be even smaller, and therefore these results justify our assumptions of spatial homogeneity of incident solar radiation.

5.4.2 Reflected Radiation in a Street Canyon and its Impact

Multiple reflections by urban materials within the urban street canyons are not considered by this GSV-based calculation method. The modeling of the contributions from these multiple reflections would require complex 3-D radiative transfer simulations. For applications on large spatial and temporal scales which require simple and fast calculations, the effect of these reflections is neglected. According to our verification results in Sect. 5.2.2, we can see that our calculation method well captures the direct and diffuse components in a street canyon. The reflectances by street buildings, trees, and roads are relatively small. The reflectance effect may become evident in the street canyons with glass walls with high reflectance at certain SZA.

The reflected radiation by urban materials (e.g., building walls, ground, and trees) exposed to direct radiation may become important when there is no direct irradiance on the street during the day, which is common in high-density street canyons during the winter season. From our results, about 9.7% of the street canyons receive zero direct solar radiation in January and about 1.0% in July (see Fig. 5.15). In this case, the radiance received is only from the diffused radiation by the atmosphere and reflected radiation by the buildings. As shown in Fig. 2 of Ali-Toudert and Mayer (2006) using 3-D numerical model simulation, the diffuse component increases as the aspect ratio H/W increases because of the reflected radiation from the buildings. The increment of about 100 W/m^2 at maximum as H/W changes from low (0.5) to high (4.0) is comparable to the total diffuse component (approximately 150 W/m^2) for the widest geometry. This indicates that in a street canyon the reflected radiation by buildings and the diffuse component by the atmosphere may be similar in terms of magnitude. The daily diffuse irradiation from the atmosphere

5.4 Discussion

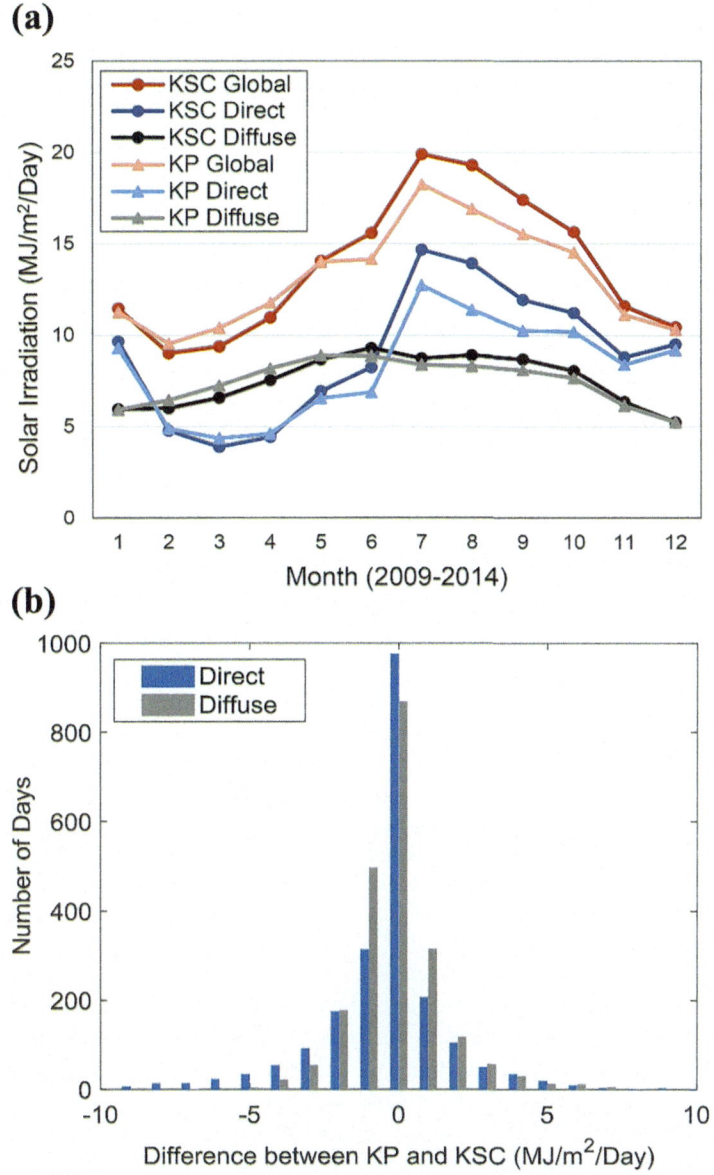

Fig. 5.14 (a) Comparison of monthly mean of daily solar irradiation measured at King's Park (KP) site and Kau Sai Chau (KSC) site of Hong Kong Observatory from 2009 to 2013; (b) The histogram of the difference between daily solar irradiation at KP and KSC

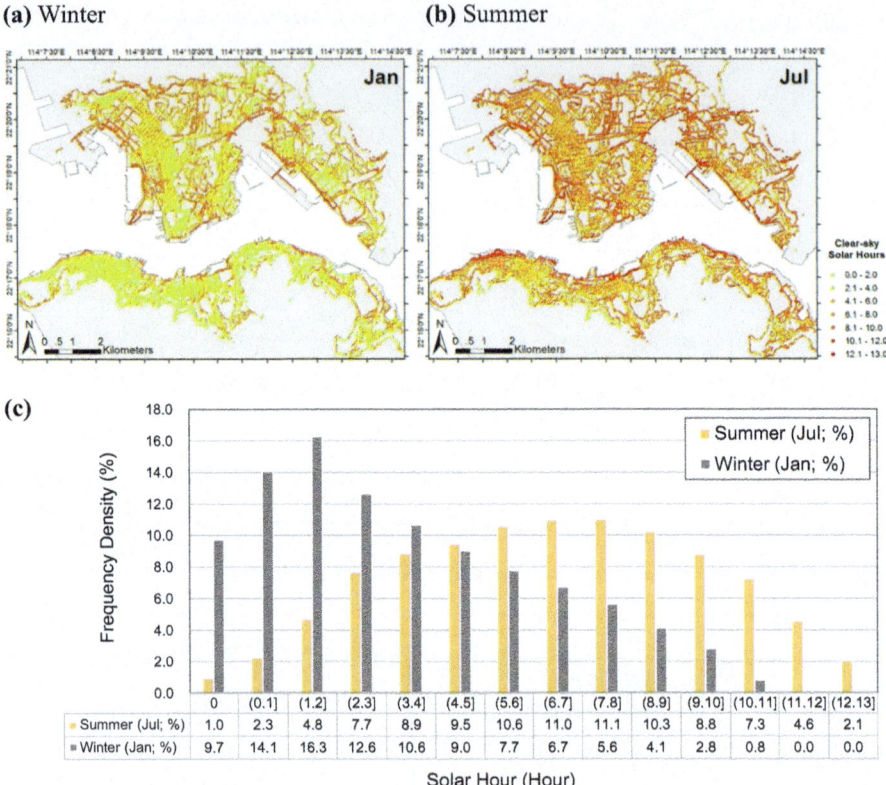

Fig. 5.15 Monthly mean of daily clear-sky solar hours in street canyons, averaged over 6 years from 2009 to 2014 in the high-density urban areas of Hong Kong, in the winter in (**a**) and the summer in (**b**). Comparison analysis of the frequency of solar hours in summer (July) and in winter (January) in (**c**)

is about 0.60 MJ/m² on average in January for the street canyons without direct solar radiation. This value is relatively small when the street canyon is exposed to direct solar radiation, which, according to our estimation when the direct solar exposure time is longer than 1 h, is larger than 3.5 MJ/m² in January (account for 73.1% of the street canyons) and 5.1 MJ/m² in July (account for 96.7% of the street canyons). Since the effect of reflections may be relatively important only when there is no direct solar radiation shining on the street in a day which only happens over a limited number of street locations, we, therefore, expect the impact of reflected radiation by buildings to be very small on the spatiotemporal patterns of daily solar irradiation. For future studies, GSV images may potentially be used to make a first-order estimation of the reflected radiation from buildings by constructing a correlation between diffuse irradiance and street view factors using 3-D simulations.

5.4.3 Transmissivity of Solar Radiation Through Tree Crowns

In this study, street trees are considered to be the same obstructions as buildings and the solar transmissivity of tree crowns is assumed to be zero. The uncertainty from this assumption may be small when trees leaves are dense. This is a reasonable assumption since Hong Kong is located in the subtropical monsoon region where the street trees can be nearly maintained throughout the year (Jim, 1987). However, it has been shown that average transmissivity of direct solar radiation through the foliated and defoliated tree crowns ranges from 1.3 to 5.3% and from 40.2 to 51.9%, respectively (Konarska et al., 2014). In the developed GSV method, this solar transmissivity ratio of street trees can be refined to a larger value based on different tree types. In cases when building surfaces overlap with tree canopies, building 3-D model may be used to extract the masked areas of trees or building surfaces. The effects of tree transmissivity on urban environments should also be investigated in future studies. Pyranometers can be used to measure crown transmissivity of local tree species, and different vegetation layer should be given different properties such as surface albedo, emissivity and transmittance. Moreover, tree transmissivity for the same tree species with different tree-canopy characteristics (e.g., leave density and canopy size) should also be investigated.

5.5 Summary

This chapter focuses on (1) quantifying the spatial and temporal patterns of street-level solar irradiation in the high-density urban areas of Hong Kong; and (2) investigating the impacts of street canyon geometries (street orientation and aspect ratio) and morphologies (sky opening and obstructions by buildings and trees) on street-level solar irradiation. Verifications of our developed method using free-horizon observatory from HKO and field measurements in a high-density street canyon show that both the clear-sky (without cloud effects) and all-sky (with cloud effects) solar irradiance of street canyons accurately capture the diurnal and seasonal cycle in high-density environments. The key points from this study are:

- A strong seasonal variation with high values in the summer and low values in the winter can be observed in both clear-sky and all-sky street-level solar radiation. The lowest averaged values are 6.6 MJ/m^2/day (December) and 4.6 MJ/m^2/day (February), and highest values are 17.3 MJ/m^2/day (June) and 10.8 MJ/m^2/day (July) for clear-sky and all-sky solar radiation, respectively. The all-sky street-level solar energy from February to June drops significantly compared to the corresponding clear-sky solar radiation because of prevalent cloud coverage over this period.
- The spatial variability of street-level solar radiation, both clear-sky and all-sky, is closely related to building densities in which much lower solar radiation is

received in streets surrounding by high-density buildings. In summer, the global radiation in an extreme low-rise region (SVF $\geq$ 0.7) on average is about three times that in an extreme high-rise region (SVF $\leq$ 0.3) and differ by about five times in winter.
- For the direct radiation, the spatial patterns are similar but the solar radiation in summer is about two times higher than that in winter. For the diffuse radiation, there are large differences in spatial patterns in the two seasons. In summer days, the areas with large SVF have higher diffuse radiation than that in winter days.
- Street orientation has a large impact on the solar radiation received by a high-density street canyon. In general, street canyons with West-East orientation receive higher solar radiation in the summer season from March to September, while lower in the remaining months. In winter when SZA is high, most solar radiation is obstructed of buildings and trees for a West-East orientation street, but the North-South orientation street is still exposed to solar radiation at noon. The impact by street orientation is larger in high-rise than low-rise street canyons.

The resulted maps of street-level solar irradiation provide crucial datasets for studying the spatial and temporal variabilities of street-level solar irradiation and understanding the interactions between solar radiation, human health and the urban thermal balance in the high-density urban environment.

References

Giridharan, R., Lau, S. S. Y., Ganesan, S., & Givoni, B. (2008). Lowering the outdoor temperature in high-rise high-density residential developments of coastal Hong Kong: The vegetation influence. *Building and Environment, 43*(10), 1583–1595. https://doi.org/10.1016/j.buildenv.2007.10.003

Gong, F.-Y. (2019). *Mapping street canyon morphology and solar radiation in high-density urban environments using street sensing approach.* The Chinese University of Hong Kong.

Gong, F.-Y., Zeng, Z.-C., Ng, E., & Norford, L. K. (2019). Spatiotemporal patterns of street-level solar radiation estimated using Google street view in a high-density urban environment. *Building and Environment, 148*, 547–566. https://doi.org/10.1016/j.buildenv.2018.10.025

Gong, F.-Y., Zeng, Z.-C., Zhang, F., Li, X., Ng, E., & Norford, L. K. (2018). Mapping sky, tree, and building view factors of street canyons in a high-density urban environment. *Building and Environment, 134*, 155–167. https://doi.org/10.1016/j.buildenv.2018.02.042

Hong Kong Observatory. (2003b). *24-hour time series of solar radiation.* Retrieved May 12, 2018, from http://www.hko.gov.hk/wxinfo/ts/display_element_solar_e.htm

Hong Kong Observatory. (2003c). *King's park meteorological station.* Retrieved March 21, 2018, from http://www.hko.gov.hk/wxinfo/aws/kpinfo.htm

Hong Kong Observatory. (2010). *Climate of Hong Kong.* Retrieved June 22, 2018, from http://www.weather.gov.hk/cis/climahk_e.htm

References

Hong Kong Observatory. (2012). *Direct and diffuse solar radiation information added to Observatory's website*. Retrieved February 22, 2018, from http://www.hko.gov.hk/press/D4/pre20100401e.htm

Hong Kong Observatory. (2018). *Climate change in Hong Kong - Cloud amount, solar radiation and evaporation*. Retrieved May 24, 2018, from http://www.hko.gov.hk/climate_change/obs_hk_cloud_e.htm

Jim, C. Y. (1987). The status and prospects of urban trees in Hong Kong. *Landscape and Urban Planning, 14*, 1–20. https://doi.org/10.1016/0169-2046(87)90002-8

Konarska, J., Lindberg, F., Larsson, A., Thorsson, S., & Holmer, B. (2014). Transmissivity of solar radiation through crowns of single urban trees—Application for outdoor thermal comfort modelling. *Theoretical and Applied Climatology, 117*(3-4), 363–376. https://doi.org/10.1007/s00704-013-1000-3

Li, D. H. W., & Lam, J. C. (2002). A study of atmospheric turbidity for Hong Kong. *Renewable Energy, 25*(1), 1–13. https://doi.org/10.1016/S0960-1481(01)00008-8

LI-COR Biosciences. (2015). *Principles of radiation measurement*. Retrieved June 13, 2018, from https://licor.app.boxenterprise.net/s/liuswfuvtqn7e9loxaut

Stewart, I. D., & Oke, T. R. (2012). Local Climate Zones for Urban Temperature Studies. *Bulletin of the American Meteorological Society, 93*(12), 1879–1900. https://doi.org/10.1175/BAMS-D-11-00019.1.

Chapter 6
Implementation of Urban Environmental Planning and Governance at Street Level

Contents

6.1	Overview...	94
6.2	Implications on Street-Level Greenery...	95
	6.2.1 Significance of TVF Map for Urban Planning Practices...........................	95
	6.2.2 Hotspots Without Street Greenery in Hong Kong...................................	96
	6.2.3 Improving Street Greenery in Hong Kong...	98
6.3	Implications on Street-Level Sky Openness...	99
	6.3.1 Significances of SVF Map for Urban Planning Practices..........................	99
	6.3.2 Hotspots with Very Low Street Openness in Hong Kong.........................	100
	6.3.3 Improving Street Openness in Hong Kong...	102
6.4	Implications on Street-Level Solar Exposure..	103
	6.4.1 Significances of Street Solar Radiation Estimates for Urban Planning..........	103
	6.4.2 Solar Under-Exposure in Winter and Over-Exposure in Summer................	104
	6.4.3 Recommendations for Urban Street Planning and Design.........................	106
6.5	Potential Applications on Urban Climatic Study and Urban Planning Practices.........	106
	6.5.1 Verification of Urban Morphology and Microclimate Models...................	106
	6.5.2 Assessment of the Feasibility for Installing Solar Panels..........................	107
	6.5.3 Estimation of Effective Albedo Based on View Factors...........................	108
	6.5.4 Impacts of Street Geometries on Urban Microclimate..............................	108
	6.5.5 Urban Planning Practices for Old and New Town Developments..............	109
	6.5.6 Comparison Analysis of Global High-Density Cities...............................	112
6.6	Summary...	114
References...		114

Abstract This chapter explores the interplay between urban morphology, microclimate, and human-centric planning in high-density Hong Kong. Using Google street view (GSV)-derived 3D metrics—sky view factor (SVF), tree view factor (TVF), and building view factor (BVF)—the study quantifies street greenery, sky openness, and solar radiation to identify critical urban hotspots. Results reveal 12.6% of street points lack greenery (TVF = 0), concentrated in high-rise, narrow corridors, while 9.7% exhibit extremely low SVF (≤0.2), obstructed by dense infrastructure. Seasonal solar analysis highlights winter under-exposure in shaded W–E oriented streets and summer over-exposure in open coastal zones. Tailored solutions include vertical greening for cramped spaces, single-side tree planting for medium-width streets, and elevated bridge vegetation. This chapter validates urban canopy models with street-level data, assesses solar panel feasibility, and compares Hong Kong's TVF (0.14) with Singapore's higher greenery (TVF = 0.26). Case studies on Mong Kok's themed streets demonstrate actionable strategies for integrating greenery and open spaces. By translating 3D morphological insights into planning practices, this work provides policymakers with evidence-based tools to enhance thermal comfort, public health, and sustainability in high-density urban environments.

Keywords Street view image · Urban morphology · Solar radiation · View factor · High-density urban planning

6.1 Overview

Characterization of urban canyon morphology and quantification of street-level energy balance are important in providing scientific evidence for the urban planning and design process at street level by policymakers. The aim of this chapter is to highlight the application and utility of this study's research findings and to provide data-driven implications on the implementations of urban planning and design at street level by summarizing the scientific knowledge on 3-D urban morphology (Gong et al., 2018; Gong, 2019) and solar radiation (Gong et al., 2019) as described in Chaps. 3, 4, and 5. In the high-density urban area of Hong Kong, the street morphology is characterized using street VFs, while the street-level solar radiation, which is the principal component that drives the street canyon energy balance, is also estimated for the study area.

The spatial and temporal patterns of street canyon morphology and street-level solar radiation should be analyzed with population density to understand their impacts from the perspective of public citizens. Figure 6.1 shows the population densities in the street block level. The extremely dense population (>100,000 persons/km^2) are distributed in all eight District Councils (DCs) and a cluster exists in West Kowloon (including Sham Shui Po and Yau Tsim Mong). As a high-density city as well as a rapidly aging society with very limited public open space, easy access to street greenery and adequate solar radiation exposure are important in promoting physical and social activities and improving public health, especially for elderly, in Hong Kong (Tang & Wang, 2007). Therefore, quantification and assessment of street greenery and solar radiation provisions are essential for providing science-based evidence in urban planning and design strategies in high-density

6.2 Implications on Street-Level Greenery

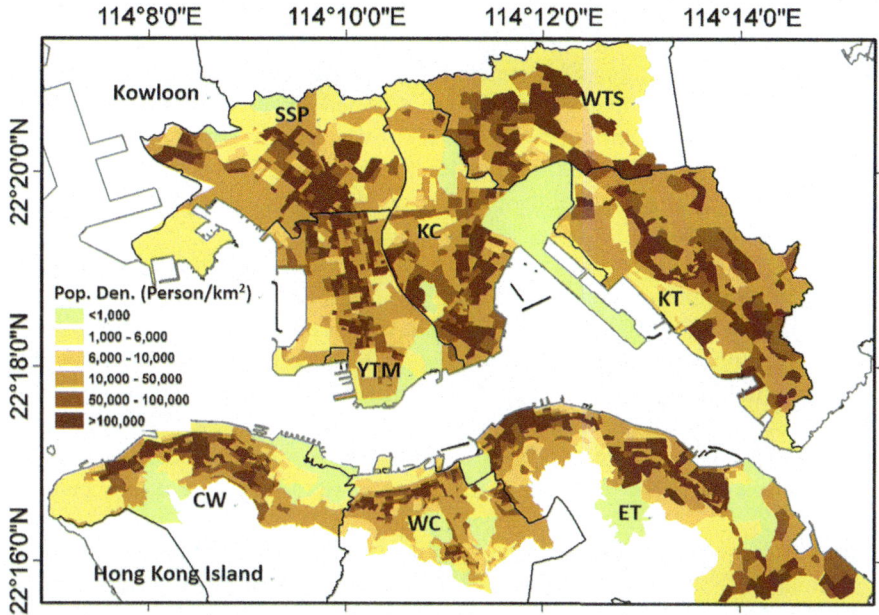

Fig. 6.1 The population density of high-density urban areas in Hong Kong based on Street Block level, including eight District Councils with five in Kowloon: Kowloon City (KC), Yau Tsim Mong (YTM), Sham Shui Po (SSP), Kwun Tong (KT), and Wong Tai Sin (WTS), and three in Hong Kong Island: Central & Western (CW), Wan Chai (WC), and Eastern (ET). The statistics of demographic are publicly available from the 2011 census in Hong Kong (Census and Statistics Department, 2016)

urban areas of Hong Kong. In this study, the high-density District Councils (DCs; the second level of planning unit), where the population density is over 6000 persons/km^2, of Hong Kong Island and Kowloon are selected as the study area, as shown in Fig. 6.1.

6.2 Implications on Street-Level Greenery

6.2.1 Significance of TVF Map for Urban Planning Practices

Access to street-level greenery is important for promoting physical activities and improving public health (Hunter et al., 2015; Schipperijn et al., 2013). Hong Kong, with meager greenery, is overwhelmingly artificial. Because of a severe space shortage, street planting is limited due to the narrow footpath, the need to cater for sightline of pedestrians and drivers, and the blockage by gigantic overhanging signboards and modern high-rise buildings. Notwithstanding the seemingly large land coverage of vegetation in Hong Kong, the built-up townscape of Hong Kong is dominated by the concrete landscape with little greenery, particularly when viewed at the street level (Hong Kong Planning Department, 2015). Planning for a greener townscape has been hindered by spatial limitations governed by urban development with greater emphasis given to population and economic growth. Planning for a greener city is therefore much needed to enhance the quality of this high-density living environment. A careful study of the physical conditions and land use characteristics of the streets are needed for improving street-level greenery.

The current statistics used in the greenery policy and most previous studies are the 2-D vegetation coverage, i.e., green coverage ratio, normalized difference vegetation index (NDVI). However, they are not representative of the realistic characterization at a street level. The existing 2-D greenery index in urban planning might not be equivalent to the 3-D greenery enjoyed by city residents from a human-scale viewpoint. The 2-D urban greenery coverage ratio might not be able to reflect an accurate exposure of greenery of people. Therefore, the 3-D quantification developed in this study is very significant for urban planning practices.

For better urban planning practices of the greening, TVF estimate has the following benefits: (1) TVF provides a realistic characterization of 3-D tree canopy at the pedestrian level and can better reflect the reality of urban environment people explored; (2) The TVF can be used in numerical model simulations to parameterize urban canopy in 3-D for urban microclimatic studies. Since tree canopy affects the calculation of street-level SVF, such incorporation of TVF in the model will improve the model characterization of SVF. In short, one main contribution of this study is that it prioritizes an actionable approach of measuring visible street greenery in 3-D which is relevant to human behaviors and experience.

As the population is growing and buildings are getting taller, the focus of street greenery planning shifts to include skyrise greenery encompassing "sky gardens," vertical planting and green roofs. The 2-D greenery coverage ratio or NDVI index may not be appropriate for street planning and design for high-density cities. Therefore, the 3-D tree canopy coverage quantified by TVF in this study is a better index to quantifying the greenery condition in cities.

6.2.2 Hotspots Without Street Greenery in Hong Kong

The "*hotspot*" for street-level greenery in this section is defined as street points without street greenery (TVF = 0). These hotspots urgently need to take some street tree planning strategies for developing a better high-density street environment. As results analysis described in Chap. 4, the average TVF value in the high-density Hong Kong is 0.14. The percentage of zero TVF value (0.0) account for 12.6% of the study area. The zero TVF is mainly limited by the high building density and narrow streets. Figure 6.2 further highlights the classification map of GSV-based TVF estimates in high-density areas of Hong Kong, in which the orange color points

6.2 Implications on Street-Level Greenery

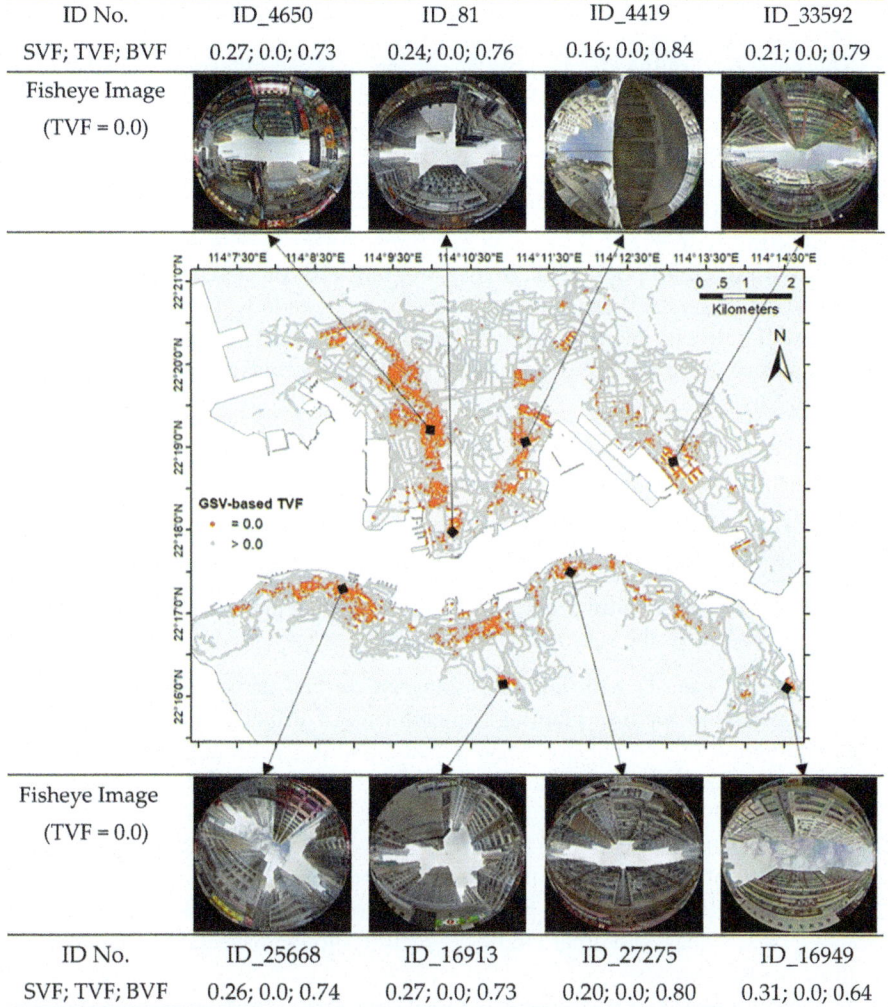

Fig. 6.2 Street hotspots without greenery (TVF = 0) in high-density urban areas of Hong Kong. The orange dots are regions with TVF equals zero, indicating no street greenery in this region. Eight examples with their fisheye images from Google Street View (with their IDs) are also shown

means the value of TVF is equal to zero, i.e., there is nearly no tree canopy or vegetation coverage in these points.

According to the highlighted hotspots and relevant fisheye images shown in Fig. 6.2, we can see that the spatial distribution of these points without street greenery mainly have the following three types:

1. The high building density, high population density, and narrow street, e.g., ID_81 and ID_16913;

2. The high building density, high population density, and medium-sized street with, e.g., ID_16949;
3. The street integrating elevated bridges or viaducts, which exposes large areas of the concrete structure, e.g., ID_4419.

Therefore, the relevant greenery planning and design strategies to improve the quality and quantity of greenery in this high-density urban environment should be proposed corresponding to three different types.

Comparing with Fig. 6.1, we can see these hotspot areas are spatially overlapping with the high population density regions. To improve the public benefit, the street greenery in these hotspot street points needs to be improved. These outputs will help city planners, health professionals, and policymakers to evaluate and improve on the tree planting in an urban street environment of high-density cities.

6.2.3 Improving Street Greenery in Hong Kong

Based on our research findings above, the corresponding planning and design strategies are proposed for the street hotspots without greenery:

1. For Street **Type (1)** with high building density and narrow street and limited space available for tree planting, other forms of urban greening, such as the incorporation of vertical greenery, including wall greening, green roof, bridge greening, and street light pillar greening, into building facades, could be considered (Jim, 1999, 2013).
2. For Street **Type (2)** with high building density and medium-size street, it is conceivable to plant single-side tree planning and consider appropriate street tree species; In addition, it is also possible to set flower buds and vines vegetation on lampposts, fences, and other landscape facilities to increase the amount of three-dimensional greenery.
3. For Street **Type (3)** with the elevated road structures, climbing plants are recommended. Flyovers inevitably have major visual impact implications, generally being unattractive and blocking view corridors and views to specific buildings. Where appropriate, mitigation measures such as using climbing plants, or other visual interest should be adopted to minimize adverse visual impact.

To improve the street greenery in the street-level hotspots, Table 6.1 provides the street cross-section and greenery improvement strategies for the three main types of the street canyon (without street greenery) in high-density urban areas Hong Kong. For three street types, vertical greening, including wall greening, rooftop greening, bridge greening, and street light pillar greening, is recommended. For the third type with relatively more street space, 3-D open green space (with ground grass, one-side street trees and vertical greenery) may be built between streets that are accessible by citizens.

6.3 Implications on Street-Level Sky Openness

Table 6.1 Greenery improvement strategies for the three main types of street canyons in high-density urban areas in Hong Kong. Both types do not have street greenery and roadside trees are not encouraged because they may block the sightline of pedestrians and drivers on streets

Aspect ratio	Street cross section	Planning strategies
H: W ≥ 3:1 (very narrow streets) TVF = 0.0 (without street tree)		• Vertical greening, including wall greening, rooftop greening, bridge greening, and street light pillar greening, may be implemented in regions with very narrow streets and no extra space for the plant
H: W = 2:1 (narrow streets) TVF = 0.0 (without street tree)		• Vertical greening may be implemented • Three-dimensional urban green spaces may be built to connect streets
H: W = 1:1 (medium-size streets) TVF = 0.0 (without street tree)		• One-side tree planning may be in consideration • Vertical greening may be implemented • Three-dimensional urban green spaces may be built to connect the streets

6.3 Implications on Street-Level Sky Openness

6.3.1 Significances of SVF Map for Urban Planning Practices

SVF is a significant factor for understanding the microthermal climate in street canyons. The SVF has been commonly used to indicate the impact of urban geometry on air temperature differences in cities. It was found that the spatial average of SVF has a close negative relationship with daytime intra-urban temperature differences (Chen et al., 2012). Moreover, SVF affects the shading level of a street (Lin et al.,

2012), the pedestrian's walking experience on a street, and solar exposure of citizens (see Sect. 6.4). In this study, GSV-based SVF estimate provides a realistic characterization of 3-D street urban environment people explored. Therefore, the SVF mapping is particularly important for street planning and design in a high-density, high-rise built-environment like Hong Kong. This chapter focuses on the street hotspots very small SVF values, which leads to insufficient solar exposure, bad ventilation conditions, and reduced street attractiveness for pedestrians.

6.3.2 Hotspots with Very Low Street Openness in Hong Kong

The *"hotspot"* for street-level sky openness in this section is defined as the street points with very low sky opening (SVF $\leq$ 0.2). For these hotspots, it is necessary to improve the street sky opening by taking some planning strategies. As the results analysis described in Chap. 4, the average SVF value in the high-density Hong Kong is 0.49. The percentage of very low SVF value ($\leq$0.2) account for 9.7% of the study area. The low SVF is mainly limited by the high-density construction and narrow streets that block sky visibility.

Figure 6.3 further shows the classification map of GSV-based SVF estimates in high-density areas of Hong Kong, in which the green color point represents the low value of SVF ($\leq$0.2) caused by dense tree canopy (TVF $\geq$ 0.3) while the red points represent the low value of SVF caused by high building density. The percentage of extreme low SVF value ($\leq$0.2) account for about 9.7% of the study area. The average SVF value of these selected points is 0.13 with a standard deviation (SD) of about 0.06. The percentage of very low SVF value caused by tree canopy account for nearly one-third.

As shown in Fig. 6.3, the very low sky opening (SVF$\leq$0.2) is located mostly at the high-density areas, especially in Hong Kong Island where there are dense tall buildings. Given the importance of street-level sky opening for the urban thermal environment and public health, the calculated sky openness hotspots based on our results suggest locations for future improvement in street-level planning and design for benefiting the general public.

According to the highlighted hotspots and relevant fisheye images shown in Fig. 6.3, we can see that the spatial distribution of these points with very low sky openness mainly have the following two types:

1. The high building density, high population density, e.g., ID_23998 and ID_4665. Some of them have horizontal overhanging signboards or elevated bridge or viaduct, e.g. ID_21505, and ID_4419. The relevant planning and design strategies to improve the street openness in this high-density urban environment should be proposed (see Sect. 6.3.3).

6.3 Implications on Street-Level Sky Openness

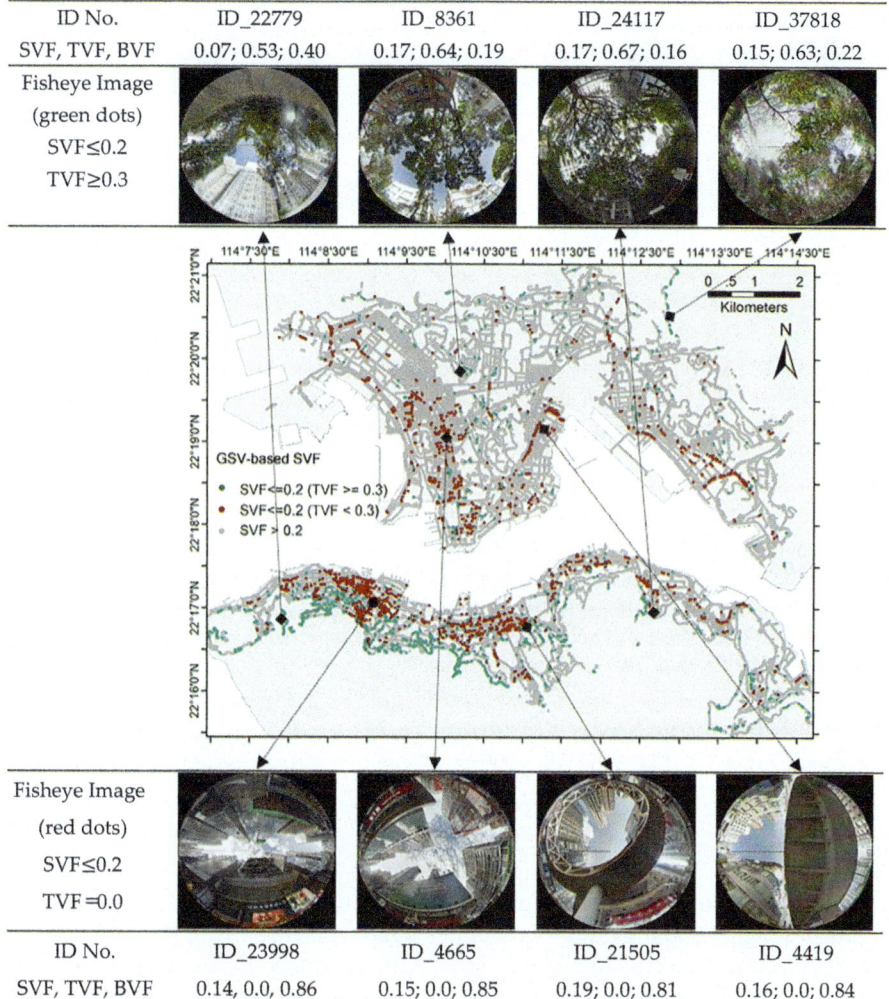

Fig. 6.3 Identifying the hotspots of very low sky openness (SVF ≤ 0.2). In particular, the green dots are regions where high tree density (TVF ≥ 0.3) block the sky openness; and the red dots are regions where high building density block the sky openness

2. The low building density, low population density, and the area covered by a large amount of tree canopy, e.g. ID_22779, ID_8361, ID_24117, and ID_37818. These regions usually have sufficient space with open sky, and therefore, are not considered in this study.

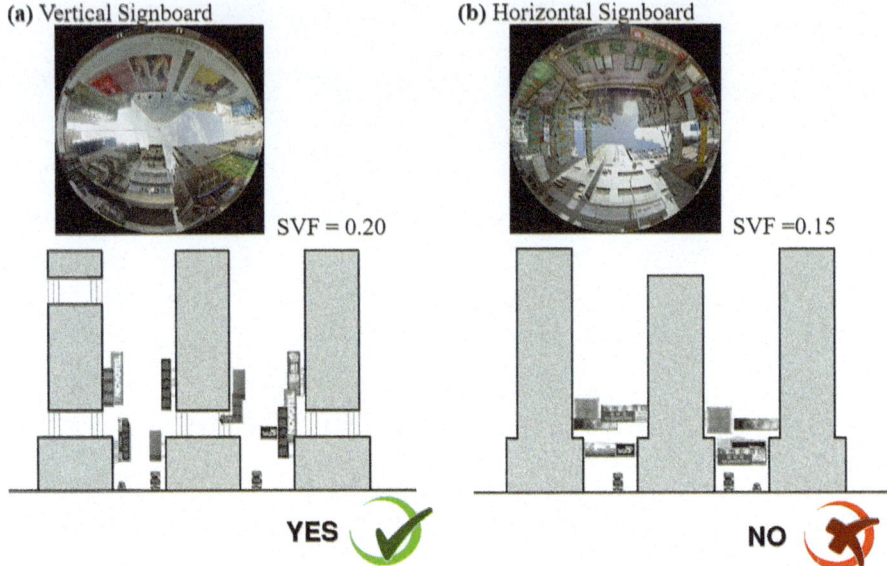

Fig. 6.4 Projecting obstructions: (**a**) Vertical overhanging signboards is benefited for increasing more openness (increasing SVF) compared with horizontal type; (**b**) Horizontal overhanging signboards should be avoided. Two realistic fisheye images and its SVF values are described

6.3.3 Improving Street Openness in Hong Kong

Based on our research findings above, the corresponding planning and design strategies are proposed as follows to improve the sky opening for streets with very low SVF: For Street Type (1) with high building density and narrow street, low SVF caused by constructions, including buildings and other projecting obstructions, such as the horizontal and gigantic overhanging signboards (see Fig. 6.4b). The projecting obstructions lead to reduce the sky opening and block the breezeways. For example, based on our results in Chap. 4, we select two points with same aspect ratio and building density, the SVF value (0.2) of the fisheye image (a) with vertical signboard is larger than that value (0.15) of the fisheye image (b) with horizontal signboards. Therefore, Signboards should preferably be of the vertical type (see Fig. 6.4a) in order to minimize wind blockage, particularly in areas with high pedestrian activities.

Secondly, as shown in Fig. 6.5, the red hotspots have very low sky openness (SVF$\leq$0.2), and the green patches are the existing open spaces. We can see that these street hotspots are located in regions where the existing urban open spaces are sparse and small in size. Especially for the three areas with black frames, are located at Mong Kok (MK), Central and Western (CW), and Wan Chai (WC) in high-density areas of Hong Kong. They have very low sky openness and lack of neighborhood open spaces. As a solution, increase the number of neighborhood urban parks, open

6.4 Implications on Street-Level Solar Exposure

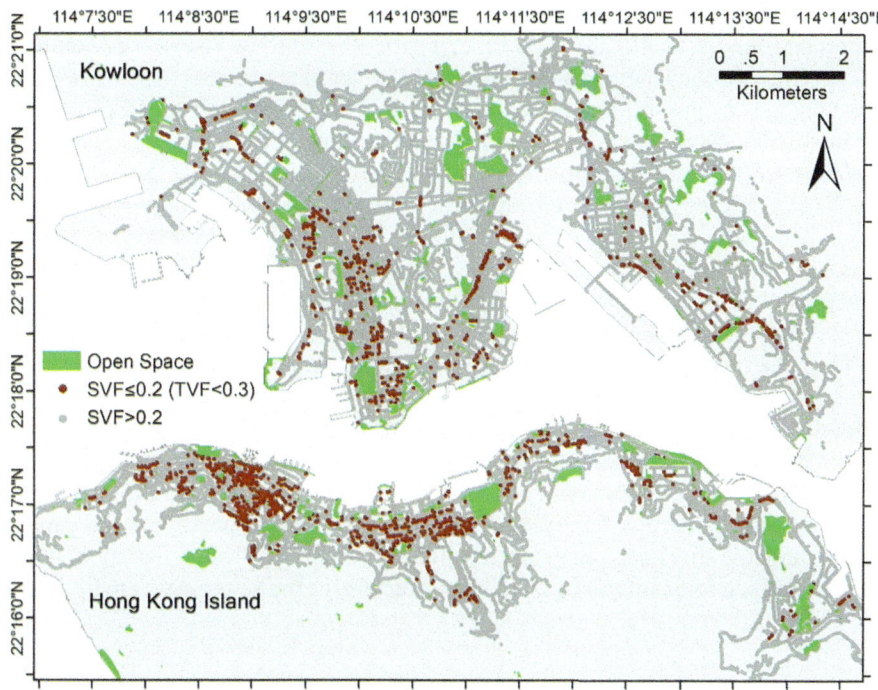

Fig. 6.5 The distribution of the urban green parks of different scales in Hong Kong. The distribution of existing urban open space, 447 patches, in the study area; The three areas marked by black frames at Mong Kok (MK), Central and Western (CW), and Wan Chai (WC) in high-density areas of Hong Kong have very low street sky openness and lack neighborhood open spaces

spaces and community parks provide more opportunities for citizens to have direct contact with open sky. In addition, the roof garden can be a good alteration to open parks or spaces.

6.4 Implications on Street-Level Solar Exposure

6.4.1 Significances of Street Solar Radiation Estimates for Urban Planning

Exposure to solar ultraviolet (UV) radiation is an important health issue for human health. On one hand, over-exposure to UV has been demonstrated in many studies to be harmful due to the diverse biological effects of UV. Among the harmful impacts on human health, skin cancer and malignant melanoma are two of the most severe ones (World Health Organization, 2006). On the other hand, insufficient-exposure to sunlight may affect the production of Vitamin D which is important for bone health and is associated with other chronic diseases if Vitamin D is low. Thus, public health policy makers and urban planners should work together to encourage an appropriate solar exposure time while preventing over-exposure for urban citizens. Hong Kong has a subtropical maritime climate with hot and humid summers while relatively warm winters. At a latitude of about 22°N, Hong Kong has the sun shining with solar elevation angles varies between a minimum value of about 44° in winter to a maximum of 90° in summer (Chong & Lee, 2014). As a result, the amount of solar radiation, in general, has peaked in July and troughs in January and February. At street level, the solar radiation is more complex due to the street canyon morphology.

The benefits of mapping street-level solar radiation in this book for better urban planning practices include: (1) solar radiation estimated using GSV images provides realistic characterization of solar exposure at street level of urban environment people explored, which help inform urban planners to identify the under-exposure and over-exposure regions that need improvement in planning and design; (2) the spatiotemporal pattern of street-level solar radiation will improve our understandings on how the street canyon geometry affect the street-level solar radiation, which will provide scientific reference for planning and design the better street environment.

6.4.2 Solar Under-Exposure in Winter and Over-Exposure in Summer

Based on the estimated solar radiation at street level in high-density urban areas of Hong Kong in Chap. 5, street orientation has a large impact on the solar radiation received by a high-density street canyon. In general, street canyons with West-East orientation receive higher solar radiation in the summer season from March to September but lower in the winter season. In winter when SZA is high, most solar radiation is obstructed of buildings and trees for a West-East orientation street, but the North-South orientation street is still exposed to solar radiation at noon. The impact by street orientation is larger in high-rise than low-rise street canyons. We further identify the insufficient-exposure (under-exposure) in winter and excessive-exposure (over-exposure) in summer as shown in Fig. 6.6. Measures on how to improve the conditions are also discussed. As shown in Fig. 6.6a, the potential insufficient-exposure regions are mostly located at the West Kowloon and most of the northern Hong Kong Island. The pattern is similar to those of building density (Fig. 3.1c) and population density (Fig. 6.1), indicating that a large population of urban citizens living in this high-density area may be suffering from insufficient solar exposure.

According to the highlighted points in Fig. 6.6a, the solar insufficient-exposure in winter are mainly distributed in regions with high building and population density areas with W–E orientation narrow street, or the street integrating elevated bridge or viaduct, as shown in "*blue*" in Fig. 6.6a. The regions with low solar radiation due to vegetation canopy are not considered here. As shown in Fig. 6.6b, compared with the population density map in Fig. 6.1, we can see that there are some

6.4 Implications on Street-Level Solar Exposure

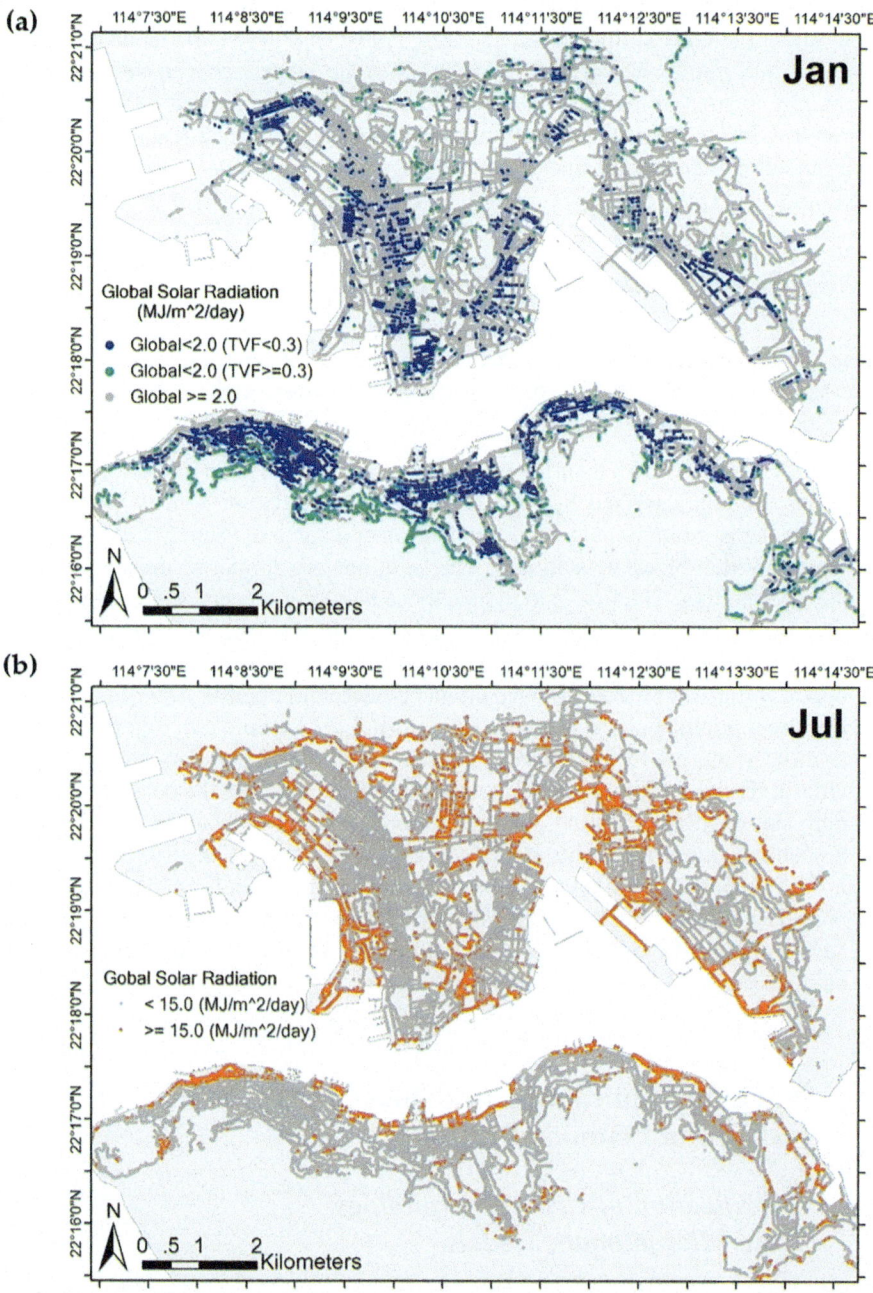

Fig. 6.6 (a) Identifying the street points of potential insufficient solar exposure in Winter (Jan.). The blue and green dots are regions at street level with daily solar radiation less than 2.0 MJ/m². This threshold is the lowest quarter of the maximum solar radiation which is about 8.0 MJ/m². In particular, the blue dots are regions with high building density while the green dots are regions with high tree density. (b) Identifying the points of potential excessive solar exposure in Summer (Jul.). The orange dots are regions at street level with daily solar radiation more than 15.0 MJ/m². This threshold is the highest quarter of the maximum solar radiation which is about 20.0 MJ/m²

regions with moderate population density in the central Kowloon as well as the areas along the coastal lines in Hong Kong Island that may experience excessive-exposure in Summer. These street points with solar excessive-exposure in summer are mainly located in regions that have the moderate building and population density areas with W–E street orientation, or open space areas outside the high-density areas and along the coastal lines.

6.4.3 Recommendations for Urban Street Planning and Design

Based on our research findings above, the corresponding planning and design strategies are proposed for regions with insufficient-exposure and excessive-exposure, respectively, as follows:

1. For streets points with insufficient solar exposure (blue plots in Fig. 6.6a). These points are characterized by high building and population density with narrow streets. Most of them have W–E orientation and some of them has elevated road structures that block the sunlight. The solutions are similar to those to improve the sky opening. The first is to reduce those blocking objects, such as huge signboards, in the streets; the second is to increase the number of neighborhood urban parks, open spaces, community parks, and roof gardens to provide more opportunities for citizens to have direct contact with open sky.
2. For street points with excessive solar exposure (orange plots in Fig. 6.6b). Some of these points are located in the moderate building and population density areas with W–E street orientation while others are located along the coastal lines with high sky opening. Planting street trees to provide more shading should be the first choice given their cooling effect. For West–East street, the tree planted at the north side of the street, which is exposed to solar radiation for a longer time, will provide necessary shade for pedestrians. Meanwhile, other shielding facilities such as pavilions and green infrastructure also need to be considered to avoid being exposed to sunshine.

6.5 Potential Applications on Urban Climatic Study and Urban Planning Practices

6.5.1 Verification of Urban Morphology and Microclimate Models

Urban models have been built to characterize the urban morphology (e.g., DSM and 3D-GIS model), urban ecosystems (Alicke et al., 1999), and urban microclimate (Bourbia & Boucheriba, 2010; Johansson, 2006). The model-based estimation method such as 3-D GIS-based and DSM-based models (Ratti & Richens, 2004; Gál et al., 2009; Chen et al., 2012) is widely used to quantify the urban morphology, such as the sky view factor and building view factor. But the accuracy is not clear because the building model may be out of date, especially in the developing worlds. The high-accuracy SVF and BVF generated from this study can be used as reference data to verify the accuracy of the model. It has been shown that in numerical simulation, if the urban morphological information is more accurate, simulations for air pollution and climate change scenarios can achieve a better result. The BVF in high-density area with little interference with street canopy will be especially valuable because it accurately quantifies the composition of a street environment. This knowledge of street composition will help the characterization of the properties of the street building surfaces.

To quantify the importance of the urban ecosystem, urban canopy model (Salamanca et al., 2010; Salamanca et al., 2011) has been made to characterize the urban trees and simulate their impact on the urban environment. However, due to its complexity in shape and structure, street tree canopy information, a major feature of urban settings, is usually very difficult to parameterize and incorporate into models. The tree view factors estimated from this study, a realistic 3-D characterization of urban trees, can be used to verify these urban canopy model, or can potentially be incorporated into these models to investigate the impact of urban canopy, such as the vertical structure of urban trees (Middel et al., 2018), on the urban environment.

Street-level solar radiation is a key component in modulating the urban thermal balance. As an important input or output of urban microclimate model (Bourbia & Boucheriba, 2010; Johansson, 2006), street-level solar radiation is usually calculated by relying on simplified assumptions on the urban geometries. Since direct solar radiation can be accurately estimated from our method (As shown in the Sects. 5.3.1 and 5.3.2), our estimation of street direct solar radiation can, therefore, be used to verify the accuracy of modeled direct solar radiation using these street-level microclimate models.

6.5.2 Assessment of the Feasibility for Installing Solar Panels

Solar panels have been an important source of household energy. To maximize the benefit, these solar panels should be installed in places with maximum solar energy potential, which is mainly determined by the solar zenith angles and building blockage in the sun's path. The smaller of the solar zenith angle and lower the blockage level, the higher the solar energy potential. In Hong Kong, as shown in Sect. 5.3, solar energy potential is the highest in the summer. On average, solar radiation is relatively steady over the course of a year in Hong Kong.

The method developed in this study can be used to evaluate the solar energy potential for a specific location. For location without horizontal blockage, the free-horizon solar radiation can be calculated using the equations in Sect. 3.4.3 by setting SVF to be one. Otherwise, if the location is lower than surrounding buildings and has a blockage in the sun's path, fisheye images can be taken and feed into our calculation model. In both cases, the height of the building should be known in order to calculate the scaled pressure at this location, for the variable p in Eq. (3.9). Global solar radiation (G_0), and its direct and diffuse components (I_{open} and D_{open}) can be conducted fort he assessment of the feasibility of installing solar panels for energy supply.

6.5.3 Estimation of Effective Albedo Based on View Factors

The effective albedo is defined as the albedo of a street canyon which is lower than the surface albedo of surrounding buildings because of the enhanced solar radiation absorption due to multiple reflections between urban structures (Oke, 1987). Understanding and quantifying the effective albedo in a city is of fundamental importance in order to estimate the solar energy absorbed by a city and simulate the local climate alteration due to urban expansion.

Our study may contribute to the calculation of the urban effective albedo in the following two ways. (1) As a realistic characterization of 3-D urban environment, street view images can be used to classify the composition of a street environment and therefore quantify the absorption capability of different building materials and eventually the absorptivity of the street canyon; (2) As shown in Morais et al. (2017), there exists an empirical correlation between the effective albedo and the SVF. The street canyon effective albedo can be estimated by the SVF. Using the estimated SVF from this study, we can give a first-order quantification of the urban effective albedo using this empirical correlation.

6.5.4 Impacts of Street Geometries on Urban Microclimate

There have been many studies investigate the impact of urban geometries in urban heat island effect, thermal comfort, and air pollution. Chen et al. (2012) found that the daytime and nighttime temperature and intra-urban temperature have a direct correlation with the sky view factor. Johansson (2006) and Krüger et al. (2011) studied the impact of an urban sky view factor on the thermal comfort of the street pedestrian. The street orientation and street sky opening, which affect the air ventilation, have also been found to affect the air pollution in a street canyon.

6.5.5 Urban Planning Practices for Old and New Town Developments

In this section, we specifically focus on the Mong Kok area where the Mong Kok Revitalisation Project is implementing by the Urban Renewal Authority (URA). URA was established in May 2001 to undertake, encourage, promote, and facilitate the urban renewal of Hong Kong based on the Urban Renewal Strategy (URS), to address the problem of urban decay and improve the living conditions in old districts. The primary goal is to improve the quality of life of residents in the urban areas by urban renewal with the concepts of sustainable development and building a quality city (URA, 2013), in which "greening" is one of the development goals. The URS has 12 main objectives for urban renewal (URA, 2011). Two of them are related to the open space and street greening as discussed in this study. They are: (1) providing more open space and community/welfare facilities; and (2) enhancing the townscape with attractive landscape and urban design. According to the Mong Kok Revitalisation Project, streetscape and pedestrian walkways improvement, and greening of the environment would be carried out to enhance the street ambience, to match and enhance the speciality of each of the five themed streets, and to highlight the variety and character of Mong Kok (URA, 2009). The five themed streets are Flower Market Road, Tung Choi Street, Sai Yee Street, Fa Yuen Street and Nelson Street in Mong Kok, as shown in Fig. 6.7. Based on the results of estimated TVF in this study, we will (1) assess the street greening of these five themed streets in this project, and (2) identify other streets in Mong Kok with a strong need of greening improvement.

Quantification of street-level 3-D greening using TVF from GSV images can help URA to identify the locations of streets with a strong need for greening improvement by planting more trees or providing more open green spaces for citizens. Mong Kok is an area with high building density, as shown in Fig. 6.6, and narrow streets, as indicated from the SVF map in Fig. 6.8a where a large portion is smaller than 0.2. From Fig. 6.8b, we can see the five themed streets (as shown in Fig. 6.6) have a certain amount of greening. Since we do not have the street view images before the revitalisation project, it is hard to make a comparison. But we

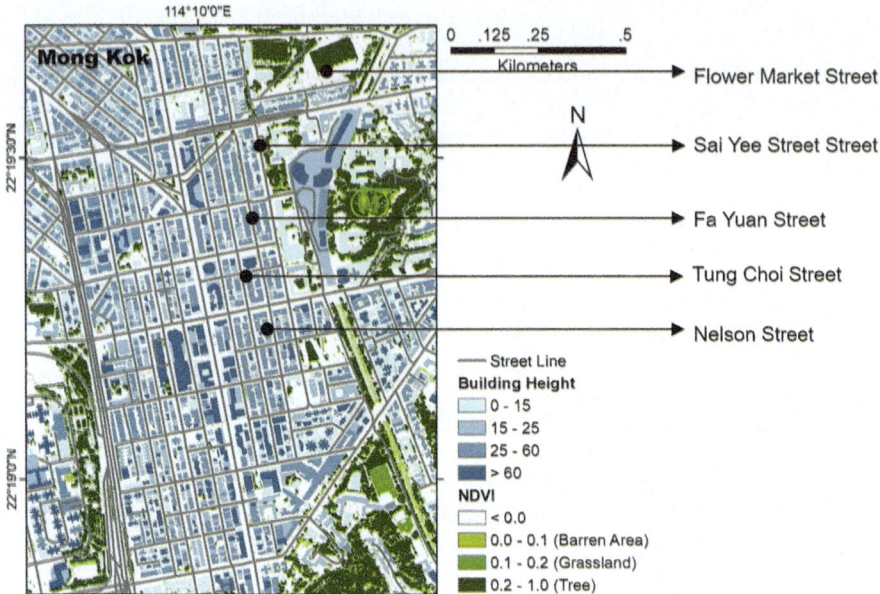

Fig. 6.7 The spatial distribution of streets, buildings, and normalized difference vegetation index (NDVI) in Mong Kok area. The (NDVI) at 1.2 m resolution is calculated using Worldview-3 data. Different vegetation type with different NDVI is also shown from grassland (light green) to trees (deep green). The five themed streets: Flower Market Road, Tung Choi Street, Sai Yee Street, Fa Yuen Street and Nelson Street from the Mong Kok Revitalisation Project are also indicated

think the street greening in these themed streets should have benefited from the project. In Fig. 6.8b, we also highlighted the regions that are lacking street greenings, as highlighted in dotted rectangle. They are the southern portions of the four streets: Sai Yeung Choi Street, Tung Choi Street, Fa Yuen Street, and Sai Yee Street. These streets should be focused in the next stage of the revitalization project in terms of improving street greening provision.

The current study has mostly focused on the potential improvements of the environment at existing urban sites. However, the methodology and principally used in this study can also be applied to the designs of new cities, such as the urban development being contemplated for artificial islands off the shores of Lantau in Hong Kong.

The methodology and principle developed in this study may help in the following way. Based on all the proposed urban planning schemes, we can (1) generate the synthetic fisheye images for the streets constructed from the urban model (with buildings and canopy) for each scheme, and (2) search for community in the real world that has similar street environment with the design scheme and create the fisheye images from Google street view image database. These data can be used to

6.5 Potential Applications on Urban Climatic Study and Urban Planning Practices 111

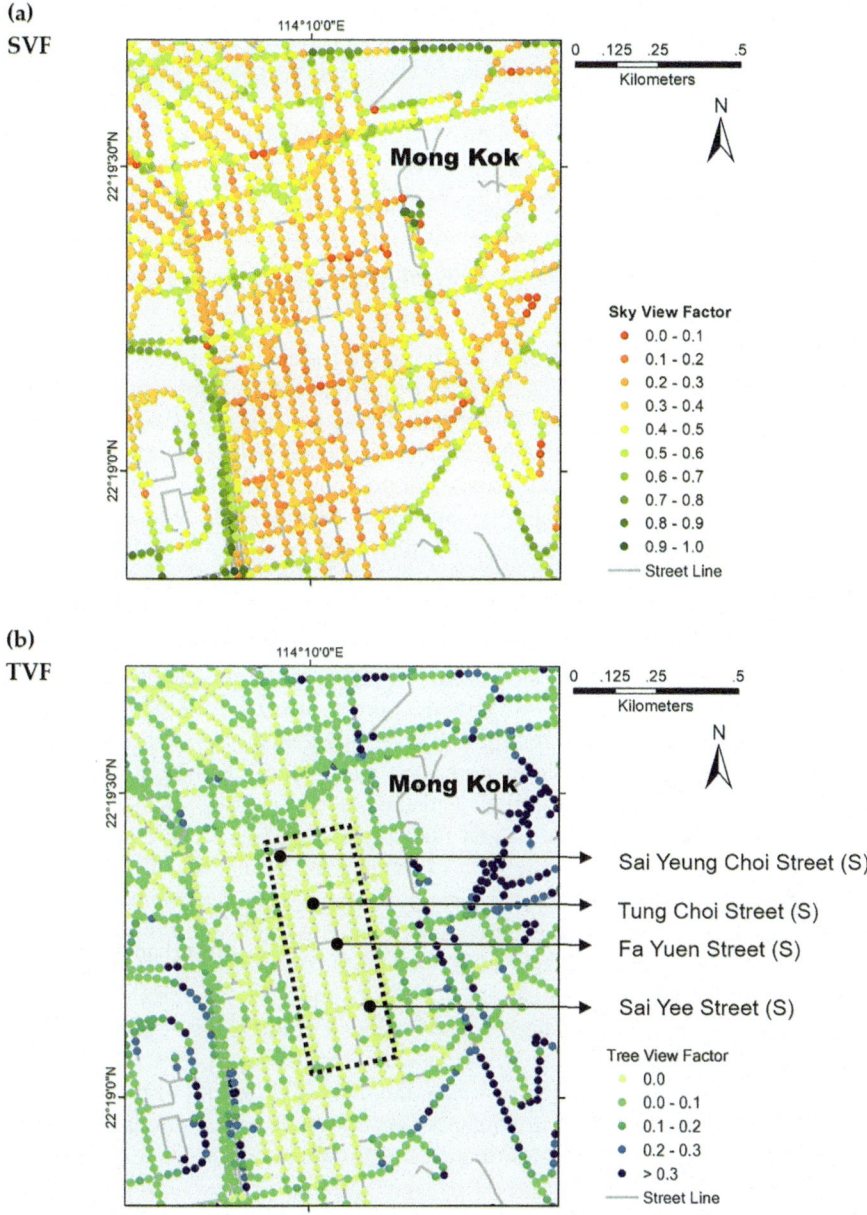

Fig. 6.8 (**a**) Map of SVF in Mong Kok and (**b**) map of TVF in Mong Kok area. The dotted rectangle highlights the area without street greening and these streets have a strong need for greening improvement. The street names are also indicated

map the VFs and solar radiation in the new town. By comparing the spatial distribution of VFs and solar radiation with the spatial distribution of residential and business areas, we can evaluate if the VFs is too low or too high, and if the solar radiation is above the over-exposure of below the under-exposure levels. In this way, the method developed in this study can provide an independent evaluation of different proposed planning scheme.

6.5.6 Comparison Analysis of Global High-Density Cities

The developed approach in this book can be applied to the global cities covered with Street View images, provided by either Google, Tencent, or Baidu, in hundreds of cities in 20 countries across four continents (Anguelov et al., 2010). The comparative analysis study of urban canyon view factors and street-level solar radiation can be conducted to explore the similarities and differences of urban morphology and thermal environment at a global scale. Such analysis would allow us to investigate the street-level greenery, sky opening, and solar radiation exposure at different cities from a public health perspective. An example in Hong Kong and Singapore can be demonstrated here.

Hong Kong and Singapore are located in a subtropical monsoon region with little effect of seasonality on the variation of the street tree canopy. A specific assumption on the seasonality is that the leaf cover of street trees does not change during different seasons even though the acquisition time of GSV images differs. Moreover, Hong Kong and Singapore are highly developed high-density city where the built-up areas are limited and therefore very little change has taken place during recent years (Census and Statistics Department, The Government of Hong Kong S A R, 2016; Government of Singapore, 2017) that will significantly affect the street skylines. Street tree canopy maps in high-density urban areas of Hong Kong and Singapore are generated in Fig. 6.9. As shown in Fig. 6.9 (a_1), (b_1), and (c_1), the SVF, TVF, and BVF values are 0.64, 0.26, and 0.08; while the relevant values in high-density urban areas are 0.53, 0.26, and 0.16. The mean TVF values in the whole Hong Kong and high-density areas of Hong Kong is 0.40, and 0.14, respectively. The mean TVF of urban areas in Hong Kong is 0.14, smaller compared with Singapore (0.26). Interestingly, the mean TVF values of the whole area of the high-density urban area in Singapore are both 0.26 with a standard deviation of about 0.20, which illustrates there is maintain medium-level greenery of whole cities.

Therefore, the comparative analysis study of urban morphologies and solar energy can be further conducted in the global cities (1) to explore the similarities and differences of urban morphology and thermal environment at a global scale;

6.5 Potential Applications on Urban Climatic Study and Urban Planning Practices

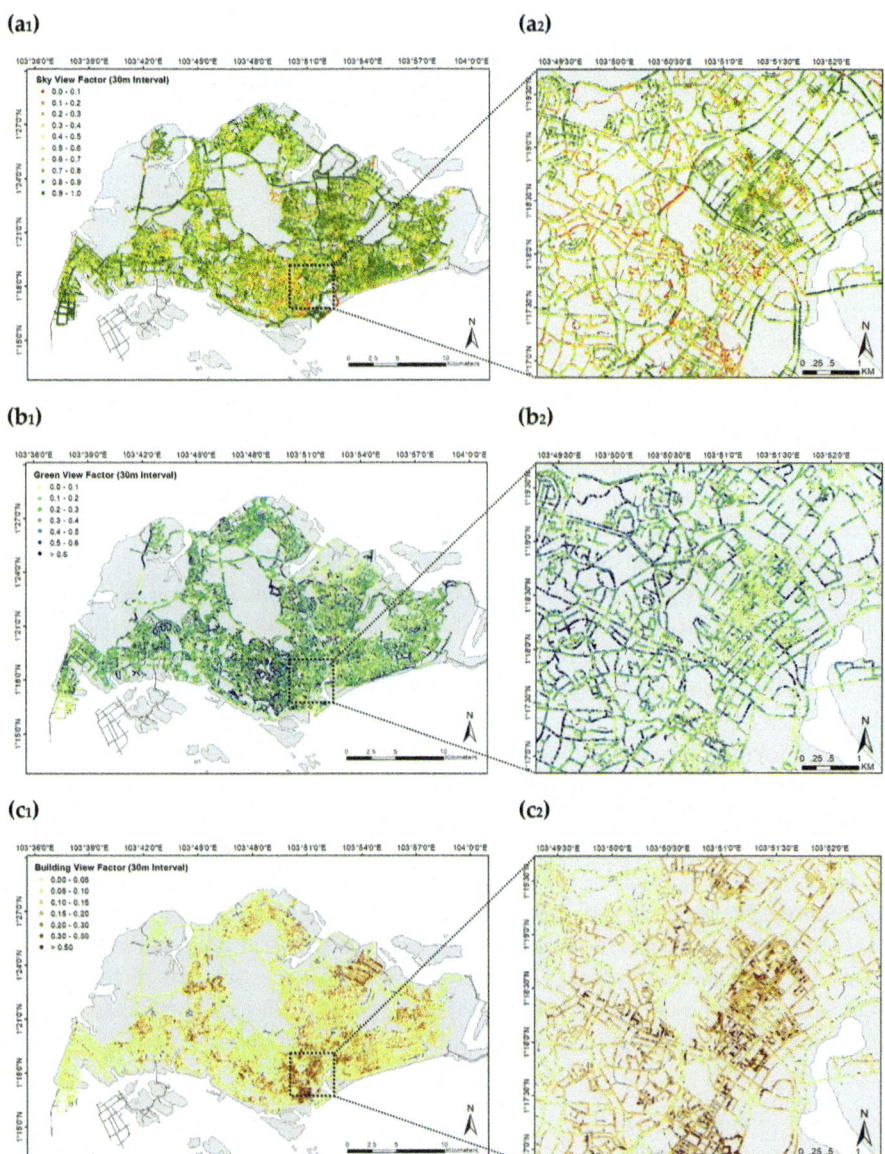

Fig. 6.9 The spatial distribution of street view factors estimates derived from 63, 488 Google Street View panoramas every 30-m interval in Singapore, including SVF in (a_1), TVF in (b_1), and BVF in (c_1); Selected areas in central Singapore [black frame in figure (**a**), (**b**), and (**c**)] in (a_2), (b_2), and (c_2)

and (2) to investigate the street-level greenery, sky openness, and solar radiation exposure at different high-density cities in public health and climatic perspectives. This study will provide policymaker, urban planner, and climatologist a data-driven implication on the implementations of urban planning and design in global cities.

6.6 Summary

This chapter integrates the findings from Chaps. 3–5, which further provides new planning and design strategies and recommendations. The new suggestions cover from implications on street-level greenery, sky openness, and solar exposure. These highlighted points illustrate the no-street-vegetation hotspots, low-street-sky openness hotspots, as well as the street areas of solar insufficient-exposure and excessive-exposure. The outputs will help city planners, health professionals, and policymakers to evaluate and improve on the tree and open space planting in an urban street environment of high-density cities.

With the results provided by this chapter, the urban planners and policymakers can first identify the local environmental issue, the hotspots without street greenery or hotspots with very low sky openness, as well as the points of solar under-exposure or over-exposure, respectively; Secondly, based on our accurate quantitative results and relevant realistic fisheye images, the study can help the researchers and planners to further identify the main types of urban streets that have these issues; Thirdly, the corresponding suggestions and planning strategies for improving street greenery, street openness, and avoiding over-exposure and under-exposure solar radiation are provided for different types of street environments.

Potential applications for (1) verification of urban morphology, ecosystem, and microclimate models; (2) Assessment of the feasibility for installing solar panels; (3) Estimation of effective albedo based on view factors; (4) Impacts of street geometries on urban microclimate; (5) Urban planning practices for new town developments; (6) Comparison analysis of global high-density cities will be considered and conducted on urban climatic study and urban planning practices.

References

Alicke, B., Hebestreit, K., Platt, U., et al. (1999). Iodine oxide in the marine boundary layer. *Nature, 397*(6716), 572–573. https://doi.org/10.1038/17508

Anguelov, D., Dulong, C., Filip, D., Frueh, C., Lafon, S., Lyon, R., & Weaver, J. (2010). Google street view: Capturing the world at street level. *Computer, 43*(6), 32–38. https://doi.org/10.1109/MC.2010.170

Bourbia, F., & Boucheriba, F. (2010). Impact of street design on urban microclimate for semi arid climate (Constantine). *Renewable Energy, 35*(2), 343–347. https://doi.org/10.1016/j.renene.2009.07.017

References

Census and Statistics Department. (2016). *2011 Hong Kong population census*. Retrieved October 31, 2018, from https://www.census2011.gov.hk/en/index.html

Census and Statistics Department, The Government of Hong Kong S A R. (2016). *Population - Overview | Census and statistics department*. Retrieved October 25, 2017, from http://www.censtatd.gov.hk/hkstat/sub/so20.jsp

Chen, L., Ng, E., An, X., Ren, C., Lee, M., Wang, U., & He, Z. (2012). Sky view factor analysis of street canyons and its implications for daytime intra-urban air temperature differentials in highrise, high-density urban areas of Hong Kong: A GIS-based simulation approach. *International Journal of Climatology, 32*(1), 121–136. https://doi.org/10.1002/joc.2243

Chong, S.-N., & Lee, T.-C. (2014). *Solar energy resources in Hong Kong from a climatological point of view*. Retrieved February 17, 2018, from http://www.hko.gov.hk/education/article_e.htm?title=ele_00443

Gál, T., Lindberg, F., & Unger, J. (2009). Computing continuous sky view factors using 3D urban raster and vector databases: Comparison and application to urban climate. *Theoretical and Applied Climatology, 95*, 111–123. https://doi.org/10.1007/s00704-007-0362-9

Gong, F.-Y. (2019). *Mapping street canyon morphology and solar radiation in high-density urban environments using street sensing approach*. The Chinese University of Hong Kong.

Gong, F.-Y., Zeng, Z.-C., Ng, E., & Norford, L. K. (2019). Spatiotemporal patterns of street-level solar radiation estimated using Google street view in a high-density urban environment. *Building and Environment, 148*, 547–566. https://doi.org/10.1016/j.buildenv.2018.10.025

Gong, F.-Y., Zeng, Z.-C., Zhang, F., Li, X., Ng, E., & Norford, L. K. (2018). Mapping sky, tree, and building view factors of street canyons in a high-density urban environment. *Building and Environment, 134*, 155–167. https://doi.org/10.1016/j.buildenv.2018.02.042

Government of Singapore. (2017). *Home | MND land use plan*. Retrieved May 21, 2018, from https://www.mnd.gov.sg/LandUsePlan/

Hong Kong Planning Department. (2015). *Planning department - Hong Kong planning standards and guidelines - Contents*. Retrieved November 2, 2018, from https://www.pland.gov.hk/pland_en/tech_doc/hkpsg/full/index.htm

Hunter, R. F., Christian, H., Veitch, J., Astell-Burt, T., Hipp, J. A., & Schipperijn, J. (2015). The impact of interventions to promote physical activity in urban green space: A systematic review and recommendations for future research. *Social Science & Medicine, 124*, 246–256. https://doi.org/10.1016/j.socscimed.2014.11.051

Jim, C. Y. (1999). A planning strategy to augment the diversity and biomass of roadside trees in urban Hong Kong. *Landscape and Urban Planning, 44*(1), 13–32. https://doi.org/10.1016/S0169-2046(98)00113-3

Jim, C. Y. (2013). Sustainable urban greening strategies for compact cities in developing and developed economies. *Urban Ecosystems, 16*(4), 741–761. https://doi.org/10.1007/s11252-012-0268-x

Johansson, E. (2006). Influence of urban geometry on outdoor thermal comfort in a hot dry climate: A study in fez, Morocco. *Building and Environment, 41*(10), 1326–1338. https://doi.org/10.1016/j.buildenv.2005.05.022

Krüger, E. L., Minella, F. O., & Rasia, F. (2011). Impact of urban geometry on outdoor thermal comfort and air quality from field measurements in Curitiba, Brazil. *Building and Environment, 46*(3), 621–634. https://doi.org/10.1016/j.buildenv.2010.09.006

Lin, T.-P., Tsai, K.-T., Hwang, R.-L., & Matzarakis, A. (2012). Quantification of the effect of thermal indices and sky view factor on park attendance. *Landscape and Urban Planning, 107*(2), 137–146. https://doi.org/10.1016/j.landurbplan.2012.05.011

Morais, M. V. B., Marciotto, E. R., Urbina Guerrero, V. V., & Freitas, E. D. (2017). Effective albedo estimates for the metropolitan area of São Paulo using empirical sky-view factors. *Urban Climate, 21*, 183–194. https://doi.org/10.1016/j.uclim.2017.06.007

Middel, A., Lukasczyk, J., Maciejewski, R., Demuzere, M., & Roth, M. (2018). Sky view factor footprints for urban climate modeling. *Urban Climate, 25*, 120–134. https://doi.org/10.1016/j.uclim.2018.05.004

Oke, T. R. (1987). *Boundary layer climates*. Routledge.

Ratti, C., & Richens, P. (2004). Raster analysis of urban form. *Environment and Planning B: Planning and Design, 31*(2), 297–309. https://doi.org/10.1068/b2665

Salamanca, F., Krpo, A., Martilli, A., & Clappier, A. (2010). A new building energy model coupled with an urban canopy parameterization for urban climate simulations—Part II: Coupling with the urban canopy parameterization. *Theoretical and Applied Climatology, 99*(3-4), 331–344. https://doi.org/10.1007/s00704-009-0142-9

Salamanca, F., Martilli, A., & Clappier, A. (2011). A new building energy model coupled with an urban canopy parameterization for urban climate simulations—Part I: Formulation, verification, and sensitivity analysis of the model. *Theoretical and Applied Climatology, 103*(1–2), 315–328. https://doi.org/10.1007/s00704-010-0302-y

Schipperijn, J., Bentsen, P., Troelsen, J., Toftager, M., & Stigsdotter, U. K. (2013). Associations between physical activity and characteristics of urban green space. *Urban Forestry & Urban Greening, 12*(1), 109–116. https://doi.org/10.1016/j.ufug.2012.12.002

Tang, U. W., & Wang, Z. S. (2007). Influences of urban forms on traffic-induced noise and air pollution: Results from a modelling system. *Environmental Modelling & Software, 22*(12), 1750–1764. https://doi.org/10.1016/j.envsoft.2007.02.003

URA. (2009). *URA initiates area-based revitalisation plan for Mong Kok*. Urban Renewal Authority. Retrieved from https://www.ura.org.hk/en/media/press-release/20090831-2

URA. (2011). *Urban renewal strategy: People first, a district-based and public participatory approach to urban renewal*. Urban Renewal Authority. Retrieved from https://www.ura.org.hk/f/page/8/4835/URS_eng_2011.pdf

URA. (2013). *Mong Kok revitalisation project*. Urban Renewal Authority. Retrieved from https://www.ura.org.hk/en/project/heritage-preservation-and-revitalisation/mong-kok-revitalisation-project

World Health Organization. (2006). *WHO | Solar ultraviolet radiation: Global burden of disease from solar ultraviolet radiation*. Retrieved November 2, 2018, from http://www.who.int/uv/publications/solaradgbd/en/

Chapter 7
Advancements and Future Directions in Urban Street Sensing Methodologies

> Now this is not the end.
> It is not even the beginning of the end.
> But it is, perhaps, the end of the beginning.
>
> —Winston Churchill

Contents

7.1	Summary of Contributions.	118
7.2	Strengths of Street Sensing Method.	120
7.3	Assumptions of Street Sensing Method.	120
	7.3.1 Spatial and Temporal Variations of Street View Factors.	120
	7.3.2 Spatial Inhomogeneity of Solar Radiation.	121
	7.3.3 Transmissivity of Solar Radiation Through Tree Crowns.	121
	7.3.4 Reflected Radiation and Its Impacts.	122
	7.3.5 Corrections for Global Measurements Under Cloudy Skies.	122
	7.3.6 Impact of Sky Luminance Distribution on Diffuse Radiation Estimation.	123
7.4	Limitations and Future Works.	124
References.		125

Abstract This chapter first highlights the summary of the contributions of this book. Then, it summarizes the strengths and assumptions of the street sensing approach in this study. Finally, it outlines the limitations and potential future works. This chapter consolidates the methodological innovations of a GSV-driven street sensing framework for urban climatic studies. By leveraging deep-learning and hemispheric photography, the approach accurately quantifies sky (SVF), tree (TVF), and building (BVF) view factors, revealing Hong Kong's high-density urban morphology dominated by low greenery and obstructed skies. Seasonal solar radiation mapping identifies winter under-exposure in shaded canyons and summer over-exposure in open zones, driven by street orientation and geometry. Strengths include global applicability via ubiquitous street view platforms and low-cost scalability. Key assumptions—stable subtropical tree cover, spatial irradiance homogeneity, and zero tree transmissivity—are justified but flagged for refinement. Limitations like ignored reflected radiation and reliance on local observatory data are discussed, with proposed solutions involving satellite integration and radiative transfer modeling. Future work emphasizes global comparative studies, dynamic sky luminance integration, and urban microclimate parameterization. This framework bridges street-level environmental assessment with sustainable urban design, offering policymakers tools to enhance thermal comfort and solar equity in high-density cities.

Keywords Street sensing methodology · Urban canyon morphology · Solar radiation modeling · High-density urban environments

7.1 Summary of Contributions

This research has made important contributions in the following two aspects:

1. This study (Gong, 2019; Gong et al., 2018) developed an effective and accurate street sensing approach, i.e., using Google Street View (GSV)-based method to accurately map all street view factors including sky, building, and tree view factors using the deep-learning technique. This approach can potentially be applicated for worldwide cities.

   (a) The proposed method accurately quantifies the openness, tree, and building patterns of street canyons in high-density urban areas of Hong Kong only using GSV images. The spatial patterns of view factor estimates are similar and consistent with the corresponding building height and density. The TVF is dominated by values less than 0.1, which is limited by the high-building density and narrow street environment.
   (b) Verification using reference data by hemispheric photography from field surveys shows that the GSV-based VF estimates have a satisfying agreement (with all R^2 values larger than 0.95) with the reference data. It demonstrates the effectiveness and high accuracy of the GSV-based method.
   (c) A comparison between GSV-based and 3D-GIS-based SVFs show that a lack of street trees in a 3D-GIS model of a street environment is the dominant factor contributing to the large discrepancies between the two datasets. This study demonstrates an effective and accurate approach for mapping SVF in high-density areas of Hong Kong and suggests that street trees should be considered in model simulations of urban street environments.
   (d) A comparison between GSV-based and 3D-GIS-based SVFs show that the two SVF estimates are correlated (with R^2 of 0.40) and have a better agreement in high-building-density areas. However, the 3D GIS-based method overestimates SVF by 0.11 on average.
   (e) The differences between the two methods are significantly correlated with street trees ($R^2 = 0.53$). The more street trees, the larger the difference (by a factor of 1.17). This suggests that a lack of street trees in a 3D-GIS model of a street environment is the dominant factor contributing to the large discrepancies between the two datasets. This study demonstrates an effective and accurate approach for mapping SVF in high-density areas of Hong Kong

7.1 Summary of Contributions

and suggests that street trees should be considered in model simulations of urban street environments.

(f) With these spatial and temporal mappings of street morphology, street-level hotspots can be identified for insufficient street-level greenery or very low sky-openness, this research provides evidence-based scientific understandings of street-level environments and corresponding suggestions for improving urban planning and design practices.

2. This study (Gong et al., 2019) further proposes an effective and accurate street sensing approach using the street geometries characterized GSV images to quantify the spatiotemporal patterns of daily and monthly mean street-level solar radiation, including global irradiance, and its direct and diffuse components.

(a) The proposed method in this study is used to map the street-level solar radiation in the urban environment and investigate its spatial and temporal pattern in high-density urban areas of Hong Kong. This method does not rely on urban digital 3-D model and therefore can be effective to calculate the street-level solar radiation. A strong seasonal variation with high values in the summer and low values in the winter can be observed in both clear-sky and all-sky street-level solar irradiation.

(b) Spatiotemporal patterns of daily and monthly street-level solar irradiation are accurately quantified. The spatial variability of street-level solar irradiation, both clear-sky and all-sky, is closely related to building densities in which much lower solar radiation is received in streets surrounding by high-density buildings. The global irradiation in low-rise regions is about three times that in high-rise regions in summer and about five times in winter.

(c) Verifications of our developed method using free-horizon observatory from HKO and field measurements in a high-density street canyon show that both the clear-sky (without cloud effects) and all-sky (with cloud effects) solar irradiance of street canyons accurately capture the diurnal and seasonal cycle in high-density environments.

(d) Direct and diffuse components of solar irradiation are quantified separately. For the direct irradiation, the spatial patterns are similar but the solar irradiation in summer is about two times higher than that in winter. For the diffuse irradiation, there are large differences in spatial patterns in the two seasons. In summer, the areas with large SVF have higher diffuse radiation than that in winter.

(e) Effect of street canyon geometry and morphology on solar radiation is further analyzed. Street orientation has a large impact on the solar radiation received by a high-density street canyon. Street canyons with West–East orientation receive higher solar radiation in the summer and lower in winter than North–South orientation. The impact by street orientation is larger in high-rise than low-rise street canyons.

(f) With the spatial and temporal mappings of street-level solar radiation, the areas of excessive or insufficient solar exposure can be identified, which provide crucial datasets for studying the interactions between solar radiation, human health and the urban thermal balance in a high-density urban environment.

7.2 Strengths of Street Sensing Method

This book proposed a street sensing method that uses the free access Google Street View image database, which represents an enormous source of information readily available for its analysis and for general use in the field of urban climatic studies. The combination of tools described here will allow the calculations of view factors and the visibility of the solar irradiance at any geographical location with Google Street View coverage.

1. Google street view image provides a direct characterization of urban streetscape, including street structures and geometry. GSV well captures the sky, building, and tree components in a street and can be used to calculate SVF, BVF, and TVF which are relevant to the urban thermal environment. GSV can also be used to quantify street geometries, one of the most important parameters for modeling street-level solar irradiances.
2. Google street view images are available in many cities all over the world, therefore, this method provides a low cost and effective streetscape mapping approach for urban studies. The street view images, from Google Maps APIs (2017a, 2017b, 2017c), Tencent Street View (TSV) APIs (2018), and Baidu Street View (BSV) APIs (2018) are publicly available covering almost all major cities in the world. Since these data products possess a fleet of vehicles equipped with similar cameras. These features make the database of street view images a source of easily comparable information to conduct comparative analyses at many geographical locations. The current coverage of these street view images surpasses that of 3-D city models, which is not always freely available.
3. Effective and accurate quantification methods for street view factors and street-level solar radiation can be built in high-density urban areas of Hong Kong. It can also be easily applied to any place in the world with available street view images. These developed methods in this book make the large-scale mapping of street canyon morphology and street-level solar radiation possible.

7.3 Assumptions of Street Sensing Method

The summary of assumptions in the street sensing approach developed in this book is as follows:

7.3.1 Spatial and Temporal Variations of Street View Factors

In this study, we use a 30-m interval for calculating VFs, assuming that this resolution would suffice to resolve the variation of VFs within a street. However, the GSV-based method is flexible in using any interval for mapping VFs of street canyons for

7.3 Assumptions of Street Sensing Method   121

study areas with different spatial scales. An assumption on the seasonality is that specifically, the leaf cover of the street tree does not change during different seasons even though the acquisition time of GSV images differs (see Fig. 4.8). This is a reasonable assumption since Hong Kong is located in the subtropical monsoon region where the street trees can be maintained throughout the year (Jim, 1989). Moreover, Hong Kong is a highly developed high-density city where the built-up areas are limited and therefore very little change has taken place during recent years (Hong Kong Planning Department, 2015) that will significantly affect the street skylines. However, for temperate climate regions, the seasonality of TVF will be a big issue, given that the street trees will be in an annual cycle of greening during growing seasons, and turning yellow and falling during the autumn and winter seasons. The developed deep-learning method in this study can be used to address the problem of VF seasonality by first training the deep-learning module with tree image samples from different seasons, and then applying to GSV images grouped into different seasons.

7.3.2 Spatial Inhomogeneity of Solar Radiation

In this study, the solar irradiance measurements in King's Park (KP) are assumed to represent the whole study area to calculate the all-sky street-level solar irradiance. Here we compare the measurements from KP and KSC sites to justify this assumption. The difference between the two characterizes the spatial homogeneity of incident solar radiation over this region. As shown the analysis results in Sect. 5.4.1, the global radiation at KP agrees closely with that in KSC. KSC, a more rural setting, has a slightly higher amount of direct solar irradiation in most months than KP in a more urban environment. The diffuse radiation is almost the same between KP and KSC, indicating the cloud diffusion effect is not causing bias in spatial distribution over this region. These results imply that the spatial difference in incident global solar radiation is small over the whole Hong Kong territory. Therefore, within the uniformly high-density urban study area, the difference should be even smaller, and therefore these results justify our assumptions of spatial homogeneity of incident solar radiation.

7.3.3 Transmissivity of Solar Radiation Through Tree Crowns

In this study, street trees are considered to be the same obstructions as buildings and the solar transmissivity of tree crowns is assumed to be zero. The uncertainty from this assumption may be small when trees leaves are dense. This is a reasonable assumption since Hong Kong is located in the subtropical monsoon region where the street trees can be nearly maintained throughout the year (Jim, 1987). In the developed GSV method, this solar transmissivity ratio of street trees can be refined to a larger value based on different tree types. As shown in the discussion of Sect. 5.4.3, the effects of tree transmissivity on urban environments should also be investigated in future studies. In cases when building surfaces overlap with tree canopies, building 3-D model, i.e., 3-D GIS model and digital surface model, may be used to extract the masked areas of trees or building surfaces (Li et al., 2018). Pyranometers can be used to measure crown transmissivity of local tree species, and different vegetation layer should be given different properties such as surface albedo, emissivity, and transmittance. Moreover, tree transmissivity for the same tree species with different tree-canopy characteristics (e.g., leave density and canopy size) should also be investigated.

7.3.4 Reflected Radiation and Its Impacts

Multiple reflections by urban materials within the urban street canyons are not considered by this GSV-based calculation method. The modeling of the contributions from these multiple reflections would require complex 3-D radiative transfer simulations. For applications on large spatial and temporal scales which require simple and fast calculations, the effect of these reflections is neglected. As discussed in Sect. 5.4.2, the reflected radiation by buildings and the diffuse component by the atmosphere in a street canyon may be relatively similar in terms of magnitude when the street canyon is exposed to direct solar radiation. Since the effect of reflections may be relatively important only when there is no direct solar radiation shining on the street in a day which only happens over a limited number of street locations, the impact of reflected radiation by buildings to be very small on the spatiotemporal patterns of daily solar irradiation. For future studies, GSV images may potentially be used to make a first-order estimation of the reflected radiation from buildings by constructing a correlation between diffuse irradiance and street view factors using 3-D simulations.

7.3.5 Corrections for Global Measurements Under Cloudy Skies

The possibility of including the dynamic effects of clouds on global irradiance. Although the models presented here were created for all-sky and clear-sky conditions, they use a reference real-time measurement of global irradiance under the free horizon and calculate the canyon irradiance relative to this. In real-life conditions, the strongest effect of clouds is the obstruction of the direct solar beam; which in turn reduces global irradiance to only the diffuse component. Algorithms can be created to automatically identify moments of the day when a cloud blocks the solar beam in a series of measurements. In order to automatically identify clouds on records of global irradiance, it may be possible to compare actual global measurements against expected cloudless-sky models. The associated uncertainties to such approximations would include momentarily enhancements of global irradiance by multi-reflection processes in clouds.

7.3.6 Impact of Sky Luminance Distribution on Diffuse Radiation Estimation

In this study, the calculation of diffuse radiation from clouds by Eqs. (3.12)–(3.14) relies on the assumption of uniform distribution of clouds, which is usually not the case. The distribution of clouds can be studied by the measurement of the sky luminance distribution. Sky luminance distribution is the luminance of the sky from all zenith and azimuth angles under different conditions. This distribution is essential to quantify the diffuse solar radiation at the street level. However, the sky luminance distributions are seldom available at the real time. There is only octas number for the fraction of clouds from HKO dataset, and the information of cloud distribution is unavailable. Therefore, in this study, the diffuse radiation from clouds under overcast, we consider the sky as uniform luminance.

The International Commission on illumination (CIE) Standard General Sky defines a set of 15 sky types with different luminance distribution. These standard skies model the sky luminance under a wide range of conditions from the heavily overcast sky to cloudless clear sky (CIE, 2003; Ng et al., 2007). Li et al. (2003) have collected sky luminance data of Hong Kong for 3 years from 1999 to 2001 and fit this measurement to the standard models. Ng et al. (2007) further proposed a reduced set to represent the dominant sky types in Hong Kong. Here, we discuss the impact of different luminance distribution on our calculation of diffuse radiation.

The luminance distribution of the 15 standard skies in June with solar altitude 85° and azimuth 80° is shown in Fig. 4 and Table 1 of Ng et al. (2007). The CIE standard sky includes overcast skies (type 1–5), partly cloudy skies (type 6–10), and clear skies (type 11–15). The diffuse radiation of CIE sky is characterized by high luminance around the sun and relatively lower luminance further from the sun. For overcast cases (1–5 sky types), the gradient is smaller compared to partly cloudy and clear cases. We can see that the uniform distribution assumption in Eq. (3.14) may not be adequate. However, we expect the bias to be small since most of the street sky open in Hong Kong is located in the middle where the luminance distribution is close to uniform. For intermediate skies, i.e., partly cloudy sky (6–10 sky types), since there is no cloud distribution available, according to Eqs. (3.12)–(3.14), the diffuse radiation is calculated to be a weighted sum of overcast cloud luminance and clear-sky luminance based on the cloud fraction. The clear sky includes the isotropic and anisotropic diffuse components. For future study, when field measurements of cloud distribution are available, we can consider the five sky types in partly cloudy condition into our estimation method. For clear skies (11–15 sky types), the luminance of the diffuse radiation closer to the sun location is stronger, which is consistent with the anisotropic solar diffuse radiation model in this study as shown in Fig. 3.9. The stronger luminance is contributed from higher anisotropic component due to aerosol scattering as described in Sect. 3.4.3 (3) Isotropic and anisotropic components of diffuse radiation.

7.4 Limitations and Future Works

The reflected radiation by urban materials (e.g., building walls, ground, and trees) exposed to direct radiation may become important when there is no direct irradiance on the street during the day, which is common in high-density street canyons during the winter season. Moreover, the intensity of reflections within canyons may vary a lot from one canyon to the other, creating complex geographical patterns. Previous studies have been using time-costly radiative transfer model to investigate the multiple reflections by urban materials within the urban street canyons which would require a complex 3-D GIS model. As a potential improvement, GSV images which provide a characterization of SVF and street features may be helpful to quantify the reflected radiation. First, a correlation between diffuse radiation and street view factors and features of buildings using 3-D simulations. Then, a first-order estimation of the building reflection can be made from the correlation. Solar irradiance is a useful parameter for modeling urban microclimates and energy budgets (Yang and Gong, 2025; Gros et al., 2011) as well as for modeling human comfort in outdoors environments (Huang et al., 2014; Lindberg et al., 2008; Matzarakis et al., 2010). Therefore, the present approaches will be further parameterized for modeling urban micrometeorological conditions. Furthermore, techniques now exist first to represent the geometry and then to include the nature of reflecting materials in urban canyons. The effective albedo based on multiple reflections from the building view factor (BVF) can be calculated and estimated as discussed in Sect. 6.5.3.

A potential limitation of the proposed methods is their dependence on local observatory measurements. The combined use of reconstructed canyon geometries and observatory measurements provided the advantage of modeling real-sky conditions for the specific canyon (i.e., including the effects of clouds on the overall solar radiation). However, any local observatory measurement will only be valid for representing local radiation within a limited geographical extent. Street canyons too far from the reference measurement will tend to diverge from those at close proximity, at least for high time resolution and variable cloud conditions. The extent to which a wider area can be represented by a local ground measurement may be a complex function of topography, cloud cover and even land use. For variable landscapes, observatory networks may be needed to create real-time realistic models. Nevertheless, if canyon irradiances cannot be represented by a local ground measurement, satellite measurements can still be used to reconstruct the reference ground level unobstructed irradiances for all sky conditions, albeit at the resolution of pixel size. For future study, the comparative analysis study of urban morphologies and solar energy can be further conducted in the global cities to explore the similarities and differences of urban morphology and thermal environment at different high-density cities in public health and climatic perspectives.

References

Baidu Street View APIs. (2018). *Baidu street view static map*. Retrieved November 4, 2018, from http://lbsyun.baidu.com/index.php?title=static

CIE. (2003). *CIE standard (CIE S 011/E:2003) Spatial distribution of daylight—CIE standard general sky*.

Gong, F.-Y. (2019). *Mapping street canyon morphology and solar radiation in high-density urban environments using street sensing approach*. The Chinese University of Hong Kong.

Gong, F.-Y., Zeng, Z.-C., Ng, E., & Norford, L. K. (2019). Spatiotemporal patterns of street-level solar radiation estimated using Google street view in a high-density urban environment. *Building and Environment, 148*, 547–566. https://doi.org/10.1016/j.buildenv.2018.10.025

Gong, F.-Y., Zeng, Z.-C., Zhang, F., Li, X., Ng, E., & Norford, L. K. (2018). Mapping sky, tree, and building view factors of street canyons in a high-density urban environment. *Building and Environment, 134*, 155–167. https://doi.org/10.1016/j.buildenv.2018.02.042

Google Maps APIs. (2017a). *Google street view image API | Google street view image API*. Retrieved October 20, 2017, from https://developers.google.com/maps/documentation/streetview/intro

Google Maps APIs. (2017b). *Street view image metadata | Google street view image API*. Retrieved November 27, 2017, from https://developers.google.com/maps/documentation/streetview/metadata

Google Maps APIs. (2017c). *Street view service | Google maps javascript API*. Retrieved March 21, 2018, from https://developers.google.com/maps/documentation/javascript/streetview

Gros, A., Bozonnet, E., & Inard, C. (2011). Modelling the radiative exchanges in urban areas: A review. *Advances in Building Energy Research, 5*(1), 163–206. https://doi.org/10.1080/17512549.2011.582353

Hong Kong Planning Department. (2015). *Planning department - Hong Kong planning standards and guidelines - Contents*. Retrieved November 2, 2018, from https://www.pland.gov.hk/pland_en/tech_doc/hkpsg/full/index.htm

Huang, J., Cedeño-Laurent, J. G., & Spengler, J. D. (2014). CityComfort+: A simulation-based method for predicting mean radiant temperature in dense urban areas. *Building and Environment, 80*, 84–95. https://doi.org/10.1016/j.buildenv.2014.05.019

Jim, C. Y. (1987). The status and prospects of urban trees in Hong Kong. *Landscape and Urban Planning, 14*, 1–20. https://doi.org/10.1016/0169-2046(87)90002-8

Jim, C. Y. (1989). Tree-canopy characteristics and urban development in Hong Kong. *Geographical Review, 79*(2), 210–225. https://doi.org/10.2307/215527

Li, D. H. W., Lau, C. C. S., & Lam, J. C. (2003). A study of 15 sky luminance patterns against Hong Kong data. *Architectural Science Review, 46*(1), 61–68. https://doi.org/10.1080/00038628.2003.9696965

Li, X., Ratti, C., & Seiferling, I. (2018). Quantifying the shade provision of street trees in urban landscape: A case study in Boston, USA, using Google street view. *Landscape and Urban Planning, 169*(Supplement C), 81–91. https://doi.org/10.1016/j.landurbplan.2017.08.011

Lindberg, F., Holmer, B., & Thorsson, S. (2008). SOLWEIG 1.0—Modelling spatial variations of 3D radiant fluxes and mean radiant temperature in complex urban settings. *International Journal of Biometeorology, 52*(7), 697–713. https://doi.org/10.1007/s00484-008-0162-7

Matzarakis, A., Rutz, F., & Mayer, H. (2010). Modelling radiation fluxes in simple and complex environments: Basics of the RayMan model. *International Journal of Biometeorology, 54*(2), 131–139. https://doi.org/10.1007/s00484-009-0261-0

Ng, E., Cheng, V., Gadi, A., Mu, J., Lee, M., & Gadi, A. (2007). Defining standard skies for Hong Kong. *Building and Environment, 42*(2), 866–876. https://doi.org/10.1016/j.buildenv.2005.10.005

Tencent Street View APIs. (2018). Tencent street view static map. Retrieved November 4, 2018, from https://lbs.qq.com/panostatic_v1/

Yang, Z., & Gong, F.-Y. (2025). Utilizing street view images to estimate solar energy potential for photovoltaic-powered buses. *Applied Geography, 177*, 103567. https://doi.org/10.1016/j.apgeog.2025.103567

Index

D
Deep learning, 5, 9, 33, 36–38, 41–43, 48, 52–55, 63, 118, 121

G
Google Street View (GSV), 5–7, 9, 10, 18, 19, 24, 25, 30, 32–44, 48, 52–60, 63–66, 70, 73, 78, 79, 82, 88, 89, 97, 104, 109, 110, 112, 113, 118–122, 124

H
High-density cities, 6, 57, 63, 94, 96, 98, 112–114, 121, 125
High-density urban areas, 5–9, 18, 19, 24, 30–34, 40, 41, 52–55, 57, 58, 60, 62–65, 77–80, 82, 85, 88, 89, 94–95, 97–99, 104, 112, 118–120
High-density urban environments, 2, 6–9, 24, 48, 52, 70, 90, 98, 100, 119
High-density urban planning, 8, 9, 94, 114

S
Sky view factor (SVF), 7–10, 16–19, 22, 24, 30, 33, 36, 38, 39, 44, 52–63, 65, 66, 70, 71, 73, 77, 79, 81, 83, 84, 89, 96, 99–102, 107–109, 111–113, 118–120, 124
Solar irradiance estimation, 71

Solar radiation, 2–4, 6–10, 16–25, 30, 39–48, 70–77, 81, 82, 84–90, 94, 103–109, 112, 114, 119–124
Solar radiation estimations, 6, 10, 30, 33, 48
Solar radiation modeling, 107, 124
Spatiotemporal variability, 90, 119
Street canyon geometry, 22, 44, 77, 82–85, 89, 104, 119
Street sensing methodology, 118–125
Street view imagery, 6, 19
Street view images, 4–6, 18, 19, 24, 30, 33, 38, 39, 41, 43, 79, 108–110, 112, 120

U
Urban canyon morphology, 9, 16–25, 94
Urban geometry, 16, 18, 25, 32, 46, 99, 107–109
Urban microclimate, 6, 17, 20, 55, 106–109, 114, 124
Urban morphology, 2, 3, 8, 16, 17, 25, 31, 94, 106–107, 112, 114, 125
Urban street canyon, 3, 6, 16, 17, 20, 44, 63, 73, 86, 122, 124
Urban thermal environments, 2–10, 16, 20, 100, 120

V
View factors, 5–10, 16, 18–20, 24, 25, 30, 33–40, 48, 49, 52, 54–55, 63, 65, 66, 71, 84, 88, 107–109, 112–114, 118, 120–122, 124

The manufacturer's authorised representative in the EU is Springer Nature Customer Service Centre GmbH, Europaplatz 3, 69115 Heidelberg, Germany. If you have any concerns regarding our products, please contact ProductSafety@springernature.com

Printed and bound by CPI Group (UK) Ltd, Croydon, CR0 4YY

26/03/2026

02078940-0013